T0397200

# ROLE OF DAMS AND RESERVOIRS IN A SUCCESSFUL ENERGY TRANSITION

Today, new and unexpected challenges arise for Europe's large array of existing dams, and fresh perspectives on the development of new projects for supporting Europe's energy transition have emerged. In this context, the 12th ICOLD European Club Symposium has been held in September 2023, in Interlaken, Switzerland. The overarching Symposium theme was on the "Role of dams and reservoirs in a successful energy transition". The articles gathered in the present book of proceedings cover the various themes developed during the Symposium:

- Dams and reservoirs for hydropower
- Dams and reservoirs for climate change adaptation
- Impact mitigation of dams and reservoirs
- How to deal with ageing dams

In conjunction with the Symposium, the 75th anniversary of the Swiss Committee on Dams offered an excellent opportunity to not only draw from the retrospective of Switzerland's extensive history of dam development, but to also reveal perspectives on the new role of dams for a reliable and affordable energy transition. These aspects are illustrated by several articles covering the various activities, challenges, and concerns of the dam community.

PROCEEDINGS OF THE 12TH ICOLD EUROPEAN CLUB SYMPOSIUM 2023 (ECS 2023, INTERLAKEN, SWITZERLAND, 5-8 SEPTEMBER 2023)

# Role of Dams and Reservoirs in a Successful Energy Transition

*Edited by*

R.M. Boes, P. Droz & R. Leroy

*Swiss Committee on Dams*

CRC Press is an imprint of the
Taylor & Francis Group, an **informa** business
A BALKEMA BOOK

Front Cover Image: David Birri

First published 2023
by CRC Press/Balkema
4 Park Square, Milton Park, Abingdon, Oxon, OX14 4RN

and by CRC Press/Balkema
2385 NW Executive Center Drive, Suite 320, Boca Raton FL 33431

*CRC Press/Balkema is an imprint of the Taylor & Francis Group, an informa business*



*Typeset by Integra Software Services Pvt. Ltd., Pondicherry, India*



*British Library Cataloguing-in-Publication Data*
*A catalogue record for this book is available from the British Library*

*Library of Congress Cataloging-in-Publication Data*
A catalog record has been requested for this book

ISBN: 978-1-032-57668-8 (hbk)
ISBN: 978-1-032-57671-8 (pbk)
ISBN: 978-1-032-44042-0 (ebk)
DOI: 10.1201/9781003440420

*Role of Dams and Reservoirs in a Successful Energy Transition – Boes, Droz & Leroy (Eds)*
*© 2023 The Editor(s), ISBN 978-1-032-57668-8*

# Table of contents

### *Theme D: How to deal with ageing dams Dam safety*

*© 2023 The Editor(s), ISBN 978-1-032-57668-8*

# Role of dams and reservoirs in a successful energy transition

The Swiss Committee on Dams is proud to host the 12th ICOLD European Club Symposium in Interlaken, Switzerland. The overarching Symposium theme is on the "Role of dams and reservoirs in a successful energy transition".

At the global scale, hydropower is by far the most important renewable energy source, accounting for one sixth of electricity generation. In Switzerland, this share amounts to almost 60%. The recent past has impressively demonstrated the importance of a reliable electricity supply for modern societies. Energy is more than electricity, but the role of electricity among the many energy carriers is and will be ever increasing, not least in the light of replacing fossil fuels with CO2-free alternatives. A recent Swiss study on nationwide risks has shown a scenario of large-scale shortage of electricity to feature the highest risk to the Swiss society, with an annual probability of occurrence of about 3% and a damage potential of several hundred billion Swiss francs. To ensure flexible storage capacity availability at all times and to limit the likelihood of a blackout in the winter 2022/23, planned refurbishment works at Swiss storage reservoirs were postponed by two years.

But dam reservoirs offer more than water for hydroelectric energy. They have also an important retention effect during floods and contribute to protect downstream infrastructure and settlements from inundations. Worldwide, irrigation and water supply are among the widespread purposes of artificial reservoirs, and even in Switzerland, a country with high annual precipitation and known to be the "Water tower of Europe", droughts increase in frequency and extent. Therefore, new reservoirs will feature multiple uses, from storage for hydropower over flood protection to water supply for domestic, industrial and agricultural use, and even artificial snow production. With 50% of European glacier ice volume located in Switzerland, the rapid glacier retreat following the substantial atmospheric warming in the Alps results in the formation of periglacial lakes forming at depressions formerly covered by glacier ice. In an attempt to adapt to climate change, new dam projects at such sites aim at artificially creating water storage to at least partly compensate for the natural water storage loss from glacier melt, which, amongst others, will lead to larger flood peaks in the downstream valleys.

The 12th ICOLD European Club Symposium coincides with the 75th anniversary of the Swiss Committee on Dams (SCD). Based on a loose union of five Swiss dam engineers, the predecessor of SCD, the Swiss Dam Commission, had been established in 1928, in the same year as the International Commission on large dams (ICOLD). On December 20, 1948, Henri Gicot chaired the founding assembly of the SCD. It comprised 68 members, many of them from construction and machine industry as well as electric utilities. With more than 300 members nowadays, of which 80 institutional members, the SCD is a private association representing the Swiss dam community within ICOLD. The Committee's main objective is to promote dam engineering including planning, construction, operation, maintenance and monitoring. To achieve this goal, SCD unites experts and specialists from various dam technology branches offering them a platform to share experiences, publish technical papers and organize symposia and workshops related to dam engineering.

We are convinced that dam and reservoir engineering keeps being vital for society and will have an important future. Several new dam projects are presently under study in Switzerland and in Europe. But in parallel, the important legacy of our predecessors requests attentive surveillance and innovative solutions to ensure the safety of a large fleet of ageing dams. In addition, climate change as well as the necessity for a better protection of ecosystems and biodiversity are challenges which dam engineers have to take up. Hence, attracting young engineers is thus an important goal of the dam community, and the ICOLD European Club Symposium is a perfect event to network and draw the attention of a broader public to the role of dams and reservoirs.

Robert Boes
President of the
Swiss Committee on Dams

Patrice Droz
Chair of the ECS2023
Organizing Committee

Raphaël Leroy
Vice-President of the
Swiss Committee on Dams

## Themes of the 12th ICOLD European Club Symposium

Theme A: Dams and reservoirs for hydropower

- Opportunity for energy generation and storage
- Large-scale storage reservoirs
- Pumped storage reservoirs
- New energy potential (PV, ...)
- Efficiency increase of existing schemes

Theme B: Dams and reservoirs for climate change adaptation

- Balancing extreme hydrological conditions (floods, droughts)
- Protection against floods
- Protection against other natural hazards (mass movements, glacier lake outburst floods, ...)
- Irrigation and water supply
- Multipurpose dams

Theme C: Impact mitigation of dams and reservoirs

- Environmental flows
- Sediment continuum
- Fish passage
- Hydropeaking
- Greenhouse gas emissions

Theme D: How to deal with ageing dams

- Dam safety
- Upgrade and refurbishment
- Extension and renewal
- Incorporating new purposes
- Decommissioning

*Role of Dams and Reservoirs in a Successful Energy Transition – Boes, Droz & Leroy (Eds)*
*© 2023 The Editor(s), ISBN 978-1-032-57668-8*

# 12th ICOLD European Club Symposium 75th Anniversary of the Swiss Committee on Dams Interlaken, Switzerland, 5-8 September

The 12th ICOLD European Club Symposium will take place in Interlaken, Switzerland on September 5th to 8th, 2023. It is an opportunity for the European members, to focus and discuss on the role of dams and the challenges that arise to adapt effectively to all changes imposed by the energy crisis, climate induced impacts, legislative evolution and ageing of dams, following the themes of the Symposium.

This event will also provide us with the opportunity of a second Board Meeting of the ICOLD European Club within the same year, to discuss additional issues in more detail.

The Symposium will also be the occasion for several European Working Groups to meet and exchange on their works regarding specific technical subjects related to penstocks and pressure shafts, internal erosion in embankment dams, dikes and levees as well as dams and earthquakes.

The Swiss Committee on Dams, founding member of the Club and one of the most active among the European Committees, is excellently committed to the organization of the Symposium, also providing several workshops and a unique opportunity for a technical visit to Spitallamm Dam construction site. Furthermore, it is important to mention that it has actively involved Swiss Committee on Dams Young Professionals in multiple organizing procedures.

The Symposium in Interlaken will also give the opportunity to celebrate the 75th anniversary of the Swiss Committee on Dams. This event will give an excellent opportunity to not only draw from the retrospective of Switzerland's extensive history of dam development, but to also reveal perspectives on the new role of dams for a safer energy transition.

Strong representation, from all member countries and sectors in dam engineering, by participating in the 12th ICOLD European Club Symposium, is highly important for the advancement and to strengthen the collaboration of the European dam community.

It is a great privilege to present the proceedings of the Symposium and I would like to congratulate the Swiss Committee on Dams and all those who contributed to this final selection.

Sera Lazaridou
President of the ICOLD European Club

*Role of Dams and Reservoirs in a Successful Energy Transition – Boes, Droz & Leroy (Eds)*
*© 2023 The Editor(s), ISBN 978-1-032-57668-8*

# Members of the 12th ICOLD European Club Symposium Scientific Committee

| | |
|---|---|
| Aufleger Markus | Austria |
| Balestra Andrea | Switzerland |
| Balissat Marc | Switzerland |
| Baumer Andrea | Switzerland |
| Boes Robert | Switzerland |
| Carstensen Dirk | Germany |
| Conrad Marco | Switzerland |
| De Cesare Giovanni | Switzerland |
| Droz Patrice | Switzerland |
| Ehlers Stefan | Switzerland |
| Erpicum Sébastien | Belgium |
| Favero Valentina | Switzerland |
| Fielding Mark | Ireland |
| Fry Jean-Jacques | France |
| Gunn Russel | Switzerland |
| Humar Nina | Slovenia |
| Jovani Arjan | Albania |
| Leroy Raphaël | Switzerland |
| Mazzà Guido | Italy |
| Mouvet Laurent | Switzerland |
| Otto Bastian | Switzerland |
| Petkovski Ljupcho | North Macedonia |
| Pinheiro Antonio | Portugal |
| Ruggeri Giovanni | Italy |
| Schleiss Anton | Switzerland |
| Sliwinski Piotr | Poland |
| Wieland Martin | Switzerland |
| Yang James | Sweden |

*Role of Dams and Reservoirs in a Successful Energy Transition – Boes, Droz & Leroy (Eds)*
*© 2023 The Editor(s), ISBN 978-1-032-57668-8*

# Sponsors of the 12th ICOLD European Club Symposium and 75th anniversary of the Swiss committee on Dams

## Main Event Partner Sponsors

## Silver Sponsors

## Bronze Sponsors

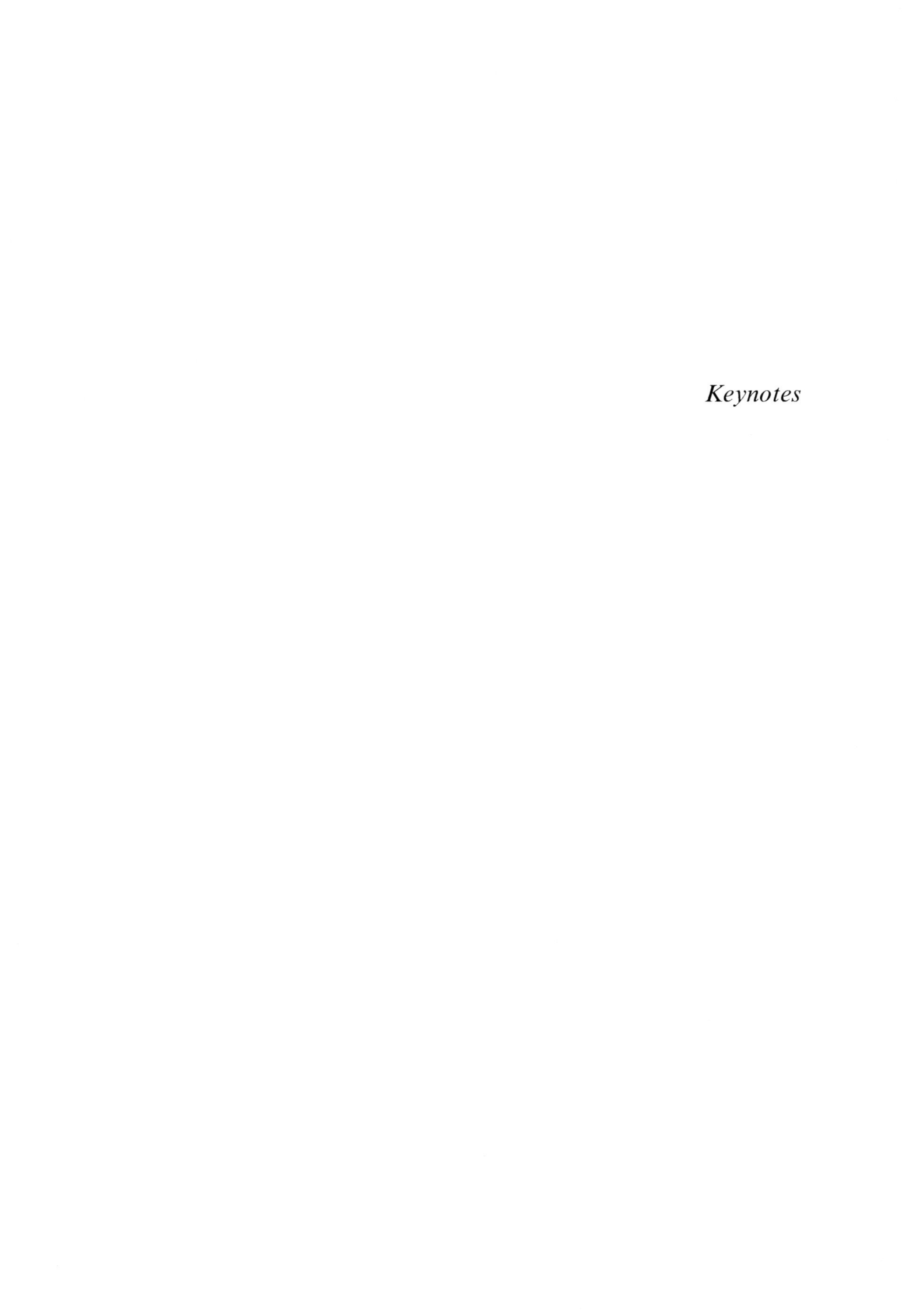

# *Keynotes*

 *ISBN 978-1-032-57668-8*

# Past, present and future role of Dams in Switzerland

R.M. Boes
*ETH Zürich, Laboratory of Hydraulics, Hydrology and Glaciology, President Swiss Committee on Dams*

A. Balestra
*Lombardi Engineering Ltd., Secretary and Treasurer Swiss Committee on Dams*

ABSTRACT: Switzerland is considered the water tower of Europe because of its topographically, and hydrologically favourable conditions with abundant water resources. The headwaters of Europe's major rivers Rhine and Rhône as well as relevant tributaries to the Danube and Po rivers, namely Inn and Ticino, respectively, are in the Swiss Alps. In lack of other major natural resources to generate electricity, the country has therefore been greatly exploiting its water resources since an early stage through the construction of storage hydropower schemes with regulating dams, now accounting for a good half of Switzerland's total annual hydropower production.

Although most of the Swiss dams were built for hydropower generation, they also increasingly provide considerable benefits as multipurpose reservoirs in terms of storage for natural hazards protection as well as agricultural, domestic, industrial and recreational scopes.

It is expected that the importance of hydropower storage on various time scales will continue to increase in the context of the envisaged renewable energy transition. Meanwhile, previously glaciated areas also offer sites for new multipurpose reservoirs. The expected challenges for Swiss dam engineering will be more and more interdisciplinary: operation and maintenance of ageing dams and hydropower plants, climate change adaptation, environmental compatibility and the increasing pressure for multipurpose exploitation of the water resources impose a comprehensive understanding and a participatory approach involving all stakeholders.

## 1 INTRODUCTION

Switzerland is a country rich in water, and dams are primarily used to store water for the generation of electricity. Larger dams were built as early as the 19th century, so that by 1928, at the time of the founding of the *International Commission on Large Dams* (ICOLD), there was already extensive national expert knowledge on dam engineering. This manifested itself in the Swiss Commission on Dams, also founded in 1928, which initially consisted of five experts who were also internationally connected and recognised. Nevertheless, it took until 1948 for the Swiss National Committee on Large Dams to be founded - later known as Swiss Committee on Dams, whose 75$^{th}$ anniversary is celebrated in 2023.

This article provides an overview of dams in Switzerland as the largest man-made structures, including the country's natural conditions, and highlights the importance of dams and reservoirs for water resources and energy management. The article concludes with an outlook on current challenges and on the importance of dams in Switzerland in the future.

## 2 FUNCTION AND PURPOSE OF DAMS

Dams, barrages and appurtenant structures are collectively referred to as dams. As the name suggests, they dam a valley considerably above the highest flood level of the natural stream. ICOLD defines a dam as "large" if its height is at least 15 m from lowest foundation to crest or a dam between 5 and 15 m with a storage volume of at least 3 $Mm^3$.

Reservoirs impounded by dams serve worldwide to store water in times of abundant inflow for periods of low inflows. In many warm and arid countries, dry periods without precipitation result in low runoffs in the watercourses, which can be compensated by releasing previously stored water from reservoirs. Occasionally, even over-year or multi-annual storage is sought, i.e. excess water of one or several wet year(s) is stored to compensate for lower precipitation and runoffs in (a) dry year(s). In the Alpine region, runoffs occur mainly in spring and summer because of precipitation in the form of rain, the melting of snow from the winter precipitation and the melting of glaciers. On the other hand, water demand for electricity production is particularly high in winter and must be compensated for by means of seasonal storage.

Dams serve worldwide a variety of purposes, primarily agricultural irrigation, hydropower, domestic and industrial water supply, low-water recharge, inland navigation and flood control. In Switzerland the focus is on hydropower utilisation, with an inherent flood protection effect.

## 3 SWISS DAM INFRASTRUCTURE

### 3.1 *Temporal development and dam types*

The oldest reservoir dam under federal supervision is the 14.5 m high earthfill dam of the Wenigerweiher on the outskirts of the city of St Gallen. It was built in 1822 to supply water and energy to nearby industrial plants and is still in operation, although kept solely for ecological reasons to maintain an amphibian spawning site of national importance (Hager, 2023).

Between 1869 and 1872, Europe's first concrete gravity dam, Maigrauge, was built at Pérolles just south of the city of Fribourg. Nearby, 50 years later, a 55 m high double curvature arch dam was built at Montsalvens, again a European premiere. This dam type later became the second most widespread of large dams in Switzerland (Figure 1). Initially, preference was given to gravity dams, such as the 110 m high gravity dam Schräh in the Wägital, completed in 1924 and considered the world's highest dam until 1929. In the years of the Great Depression and during the Second World War, some material-saving buttress dams were built, such as the Lucendro dam on the Gotthard Pass. More information on the main phases of Swiss dam development is given by Pougatsch & Schleiss (2023).

The economic upswing after the Second World War called for more hydropower, which led to a vigorous construction of dams in Switzerland: In the short period from 1950 to 1970, around 90 large dams were built (Figure 1). This corresponds to almost half of all large Swiss dams existing today. The Grande Dixence dam, which at 285 m is still the highest concrete gravity dam in the world, was also built during these years. From the 1980s onwards, the construction of large new dams in Switzerland slowed down, while the construction of small dams with other purposes than hydropower, mainly for flood retention, continued (Schwager et al., 2023). Given their knowledge and experience, Swiss engineers have been and continue to be involved in numerous dam projects abroad (Droz, 2023). In doing so, they keep pace with the constantly growing level of experience, which in turn benefits the optimal monitoring and adaptation of domestic dams and sets the basis for new Swiss dam development and refurbishment projects.

Today, there are 200 large dams in Switzerland according to the ICOLD definition (Figure 1). More than 220 dams, weirs and corresponding reservoirs are under the overall supervision of the Swiss federal authority (Schwager et al., 2023). These essentially include dams which either have a storage height of more than 10 m, or a storage height of at least 5 m and a storage volume of at least 50'000 $m^3$. The relevant storage height is defined as the

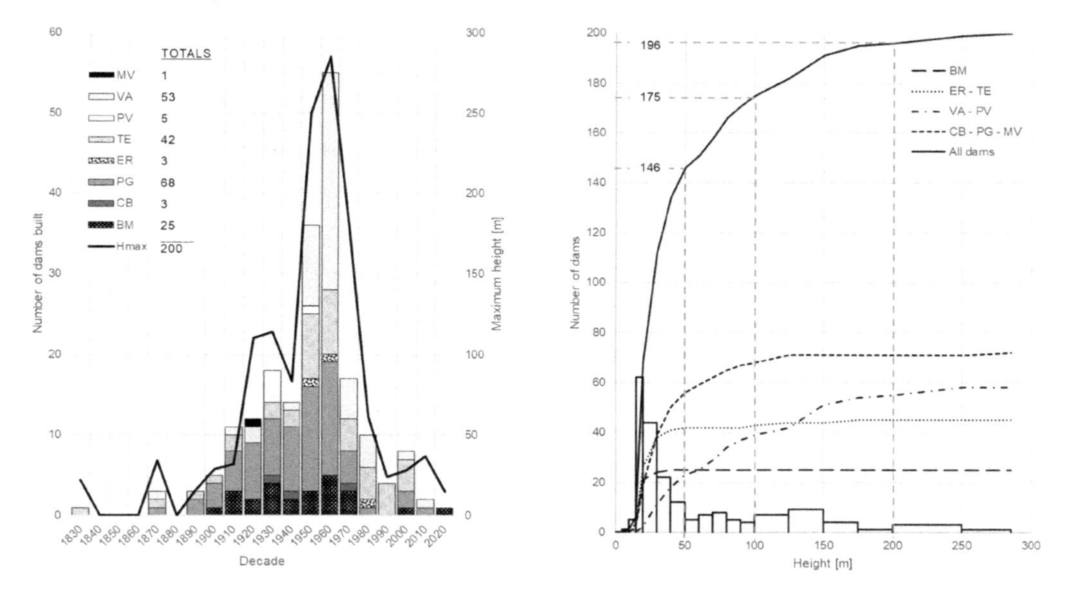


Figure 1. Development of large dams in Switzerland by type (left) and by maximum height above foundation (right); classification according to ICOLD: BM = barrage, CB = buttress dam, ER = rockfill dam, MV = multiple arch dam, PG = gravity dam, PV = arch gravity dam, TE= earthfill dam, VA= arch dam (own representation).

vertical distance between the full supply level and a lower reference level, the latter corresponding to the low water level or the ground elevation. Most large Swiss dams are located in the Alps (Figure 2).

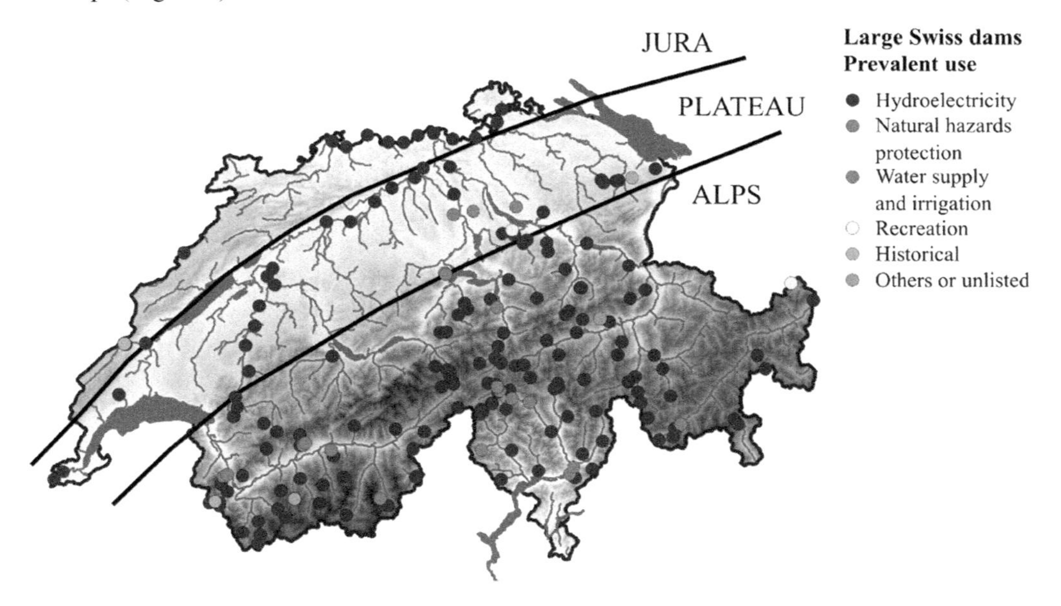


Figure 2. Spatial distribution and purposes of large dams in Switzerland (own representation).

Although well-maintained dam structures can fulfil their functions for decades, dam ageing nevertheless makes great and increasing demands on monitoring, maintenance, and operation. Negative phenomena that occur or are noticed only after a long operation time are, for example, concrete decay processes such as swelling (SCD, 2017a). The maintenance of the existing dam fleet is a demanding challenge for the specialists involved and is often technically supported by the Swiss federal institutes of technology in Zürich and Lausanne as well as other higher education institutions. As of 2023, 165 out of 200 Swiss large dams are older than 50 years, while the average age of the large Swiss dams is 70 years (Figure 3).

The demand for new dams has declined sharply since the millennium change but has recently been the subject of intense discussion again (Felix et al., 2022; Fauriel et al., 2023), see also sections 6.1 and 6.2. Heightening of existing dams is one way to increase the storage volume for winter energy production. In the past, for example, the Mauvoisin (Canton Valais) and Luzzone (Canton Ticino) arch dams were raised by 13.5 and 17 m, respectively, for this purpose (Schenk & Feuz, 1992; Baumer, 2012). An overview of the dam heightening experience of Swiss engineers is provided by Wohnlich et al. (2023).

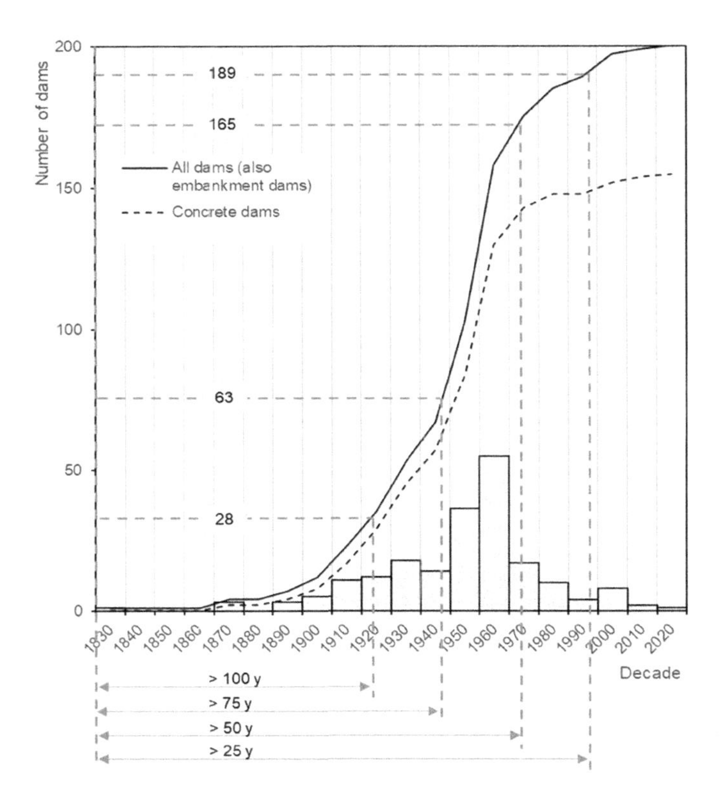


Figure 3. Development of large dams in Switzerland by age (own representation).

## 3.2 *Dams for hydropower*

Thanks to the favourable natural conditions regarding topography, hydrology and geology, the hydropower potential in Switzerland has been intensively exploited since the first power plant for electricity production in 1878 in St. Moritz (St. Moritz Energie 2023) until today. There are more than 650 run-of-river, storage, and pumped-storage power plants with installed capacities ranging from 300 kW to the 1'269 MW of the Bieudron hydropower plant (HPP). Their annual production is nominally around 36 TWh/a, which corresponds to about 57% of the country's electricity demand. In an international comparison, Switzerland has by far the highest hydropower generation per unit land area.

The following two main types of HPPs are distinguished in Switzerland:

- Low-head (below 30 m), run-of-river HPPs on the major rivers,
- High-head (200-2'200 m) HPPs with storage dams mostly of medium to large height, generating in one, two or three stages on Alpine tributaries.

Between these two predominant groups, the medium head (30-200 m) range is only sparsely represented in Switzerland, although the corresponding type of HPP is very common worldwide and also constitutes the largest HPPs, i.e. on water-rich streams with relatively high dams, and a powerhouse located nearby. The corresponding reservoirs are characterized by a small to medium capacity inflow ratio (e.g. smaller 0.25), that still allows the HPP to be operated largely independent of fluctuations in inflow with only moderate fluctuations in reservoir water level to meet demand.

The technical potential for hydropower development in Switzerland is often quantified as being exploited by around 90%, so that the potential for further expansion by construction of new schemes is rather low. Current efforts focus on maintaining and optimizing the capacity and efficiency of existing HPPs and related dams. In the future, the value of hydropower is expected to lie even more in storage on various time scales (short-term to seasonal) and the provision of great flexibility for load absorption and short-term generation by means of pumped storage (Boes et al., 2021). Pumped-storage power plants (PSPs) can be used either for electricity production or for pumping operation over the same gross head. No energy is generated in this cyclic process, as the availability of energy is only shifted over time. However, PSPs are of great value in large-scale electricity grids, as they allow to store excess energy in the form of water that can be released to generate electricity within tens of seconds to minutes, with a cyclic or round-trip efficiency of 70 to 80 %. Swiss PSPs are typically designed to store excess natural inflows in seasonal, high-head reservoirs. On the other hand, some storage HPPs also feature pumps to transfer water from lower-lying stream intakes to a reservoir. The largest Swiss storage scheme at Grande Dixence in the Canton of Valais comprises four large-scale pumping stations (Leroy, 2023).

### 3.3 *Dams for other purposes*

Alongside hydropower, the large seasonal reservoirs in the Alps contribute considerably to flood retention and hence downstream flood protection. However, important for the protection of the population from natural hazard in various regions of Switzerland are also sediment retention and flood retention dams. In addition, numerous small reservoirs (ponds) have been created in the Alps in recent decades, which serve for artificial snowmaking on ski slopes. Especially in the Canton of Valais there are several reservoirs serving for irrigation and water supply. Finally, reservoirs that no longer serve a purpose of economic relevance have ecological and recreational functions nowadays (Figure 2), so that dam removal has not yet been a topic for large Swiss dams. On the contrary, a few mostly small reservoirs impounded by dams are kept in function to preserve habitats that have established in the course of time, as exemplified by the Wenigerweiher in St Gallen (see section 3.1).

## 4 WATER TOWER SWITZERLAND

### 4.1 *Topography and relief*

With an area of 41'285 km$^2$, Switzerland is one of the smaller countries in Europe. The total extension in east-west direction is about 350 km, and in north-south direction about 220 km. Switzerland is very mountainous by nature and can be divided into three major landscape areas, which show great differences: the Jura, the densely populated Plateau and the Alps (Figure 2). Around 60 % of the country's area belongs to the Alps, 30 % to the Plateau, and 10 % to the Jura. Switzerland has more than 3350 peaks above 2'000 m a.s.l. The sixteen highest peaks in Switzerland are all in the Valais Alps and the highest peak is the Dufourspitze (4'634 m a.s.l.) in the Monte Rosa massif close to Zermatt.

## 4.2 *Climatic regions and Hydrology*

The Alps form an important climate and water divide in Central Europe with additional Alpine and intra-Alpine weather effects, so that several weather situations prevail in Switzerland despite its small size, providing a great variability in precipitation distribution (Figure 4). North of the Alps, there is a temperate, Central European climate mostly dominated by oceanic winds, whereas south of the Alps it tends to be more Mediterranean.

The mean annual precipitation is about 1400 mm (reference period 1981 to 2010). In Alpine regions, the mean precipitation is higher, e.g. 1750 mm per year, depending on the elevation. In the Plateau and the Jura, the amount is about 1000 to 1500 mm per year. Precipitation in Switzerland is about twice as high in summer as in winter. Around 32 % of Switzerland's annual precipitation evaporates, with the remaining 68 % flowing abroad as runoff (Viviroli & Weingartner, 2004).

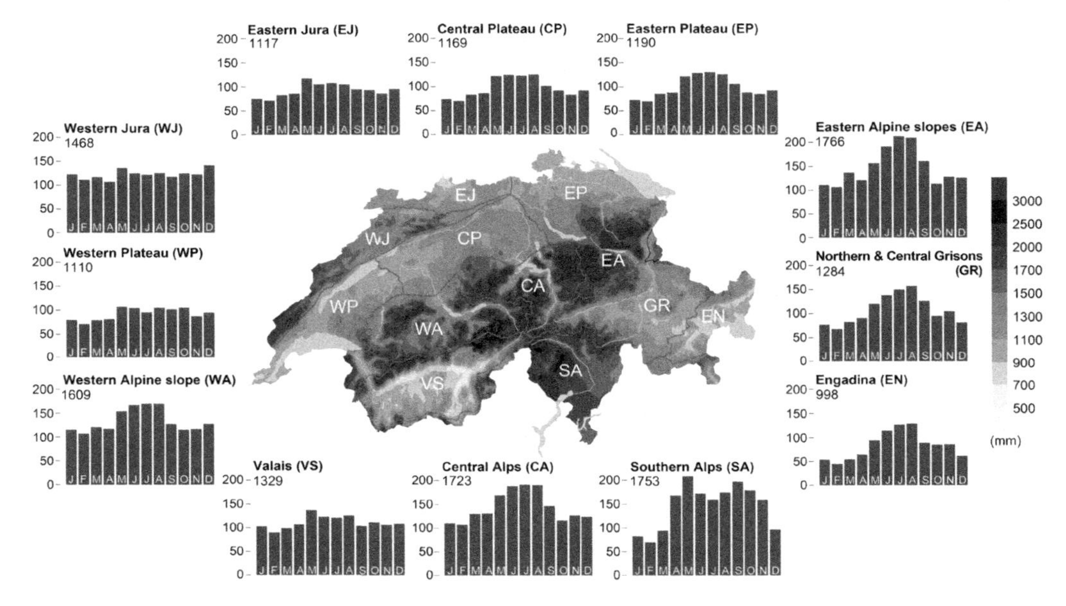


Figure 4. Annual mean precipitation and monthly sums [mm] of the measurement series 1981 to 2010 for the twelve Swiss climate regions; the annual means are indicated in the upper left corners of each barplot (Source: adapted from CH2018, 2018).

## 4.3 *Glaciers*

Swiss Alps are significantly shaped by numerous glaciers. The largest and longest Alpine glacier is the Great Aletsch Glacier, followed by the Gorner Glacier (by area). The Swiss glaciers reached their last peak during the Little Ice Age, which lasted from the beginning of the 15th to the middle of the 19th century. Since the end of the Little Ice Age, a clear retreat of the glaciers has also been observed in Switzerland, as almost worldwide. This glacier retreat has intensified in recent decades. Between 1920 and 2019, the ice volume of all glaciers in the Swiss Alps decreased by about half to 51 km$^3$, of which 38% alone occurred between 1980 and 2019 (Zekolari et al., 2019).

## 4.4 *Climate change impacts*

Due to the global temperature increase, the runoff distribution over the year is changing and the amount of stored water in snow and glacier ice will further decrease, thus affecting the multipurpose exploitation of the resource. Figure 5 shows for typical catchments and discharge regimes how seasonal runoffs will change over the course of the year until the middle and end of the century, provided that no climate protection measures are taken, and emissions and warming continue to increase (emission scenario RCP8.5).

By the end of the century, an average increase in winter runoff of around 10 % is expected with climate protection actions (RCP2.6) and around 30 % without such actions (RCP8.5) (Brunner et al. 2019). Winter runoffs increase particularly strongly in today's nival, i.e. snowmelt-dominated regimes (see e.g. Figure 5, Plessur). The smallest changes in winter runoffs are expected in catchments in the Plateau, where snow cover already contributes little to runoff, and in very high catchments (Figure 5, Rosegbach), where most precipitation will continue to fall in the form of snow in the future due to low winter temperatures (Mülchi et al. 2021).

By the end of the century, the summer scenarios show an average decrease in runoff of around 10 % with climate protection actions and 40 % without. Responsible for this decrease are reduced summer precipitation, higher evaporation and the reduction of glacier and snow melt water. Areas of all altitudes and regions are affected by declining summer runoff. There will also be a decrease in summer runoffs in areas that are still glaciated today (Figure 5, Rosegbach).

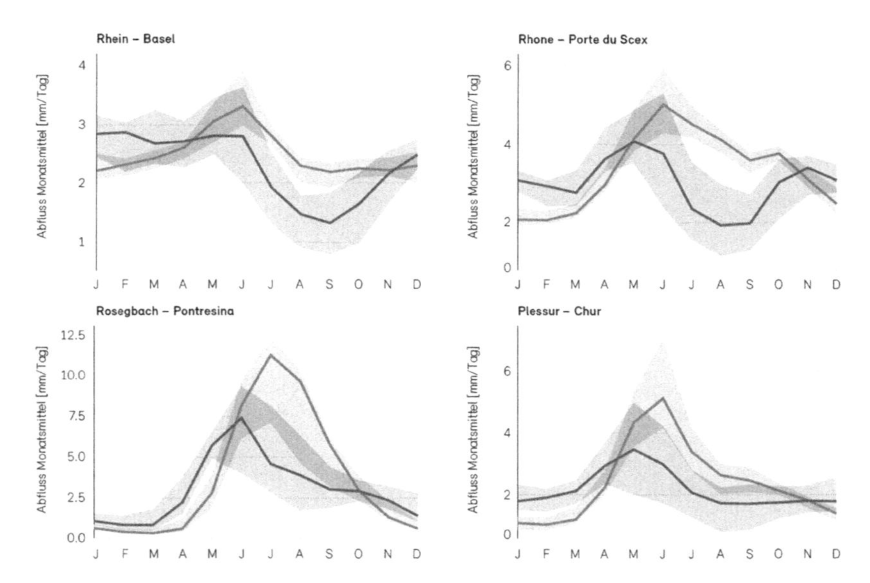


Figure 5. Change in mean monthly runoffs in characteristic catchments with median (line) and uncertainty range (fill) for the reference period (1981 - 2010) (grey) and the scenario without climate protection actions (RCP8.5) by the end of the century (2085, brown line and fill) (Source: adapted from FOEN, 2021).

## 5 SYSTEMIC RELEVANCE OF SWISS DAMS FOR BALANCING WATER RESOURCES

### 5.1 *Swiss electricity system*

Water resources in Switzerland are abundant, but the runoff regimes are subject to large seasonal fluctuations, particularly in small catchments at high altitude. In winter, the runoff is lower than in summer while the electricity demand is higher (55 % of the annual demand in the decade 2013 to 2022). To counteract this imbalance, seasonal hydropower storage systems play a relevant role.

For rivers with small catchments in the side valleys of the Central Alps, the balancing of power generation in summer and winter requires a storage volume of typically 30-40 % of the annual runoff, depending on the degree of glacier coverage in the catchments. For the rivers of the main valleys, this share is significantly lower, but in absolute terms very large storage volumes would be required. In this respect, the natural sub-Alpine between the Alps and the Plateau have an important balancing function, partially decoupling the catchment areas above and below thanks to their retention effect.

In the large hydropower reservoirs in Switzerland, a considerable part of the natural inflows is transferred from the summer to the winter half-year (1st October to 31st March), so that today the storage power plants generate half of their annual production in the summer and half in the winter half-year, whereas the inflows are distributed over these two half-years at a ratio of about 4:1.

The ratio of electricity production of storage and run-of-river HPPs in Switzerland is approximately 53 %: 47 %, while globally it is around 33 %: 67 %. Almost 90 % of the total storage capacity of Swiss hydropower reservoirs of around 9 TWh is provided by reservoirs with volumes above 20 $Mm^3$ Figure 6). This storage capacity represents a valuable reserve that renews itself every year and that increases the security of electricity supply, especially in the winter half-year. The water stored in the reservoirs until late summer is sufficient to cover around 25 % of the country's total electricity demand in the winter half-year.

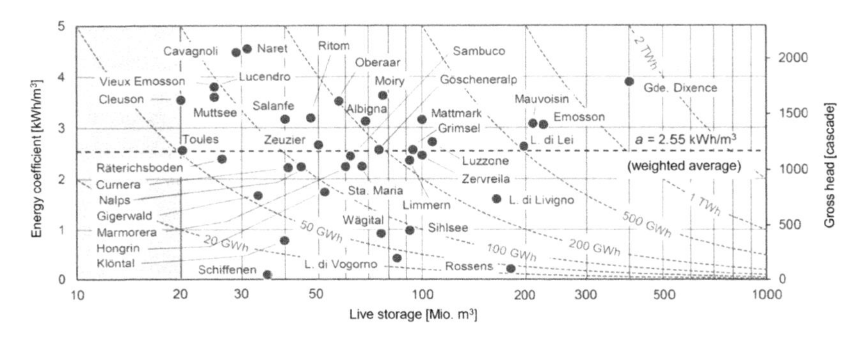


Figure 6. Storable electric energy (dashed curves), average exploitable gross head (of the HPP cascade if applicable) and corresponding energy coefficient of Swiss reservoirs above 20 $Mm^3$ as a function of their live storage volume (adapted from Felix et al., 2020).

## 5.2 *Flood risk management*

Natural and artificial lakes can significantly contribute to prevent downstream flood damage through their water retention effect. Every lake has a retention effect because inflow is always delayed and attenuated, even when fully filled. For dams and weirs, this depends on the capacity and mode of operation of the spillway, i.e. the lake outflow. The larger the lake area, the greater the attenuation effect. Thus, flood risk management is increasingly considered in the operation of the relevant Swiss reservoir dams, retention basins and regulated natural lakes.

After the 1993 flood in Valais, the spillway of the Mattmark reservoir was converted in such a way that the uppermost 2 m of the reservoir with a volume of 3.6 $Mm^3$ serve exclusively for flood retention, which the canton of Valais has at its disposal (Biedermann et al., 1996; Sander & Haefliger, 2002).

The Sihl valley in the cantons of Schwyz and Zurich has experienced no major flooding since the construction of a hydropower dam on the Sihl river some 40 km upstream of the city of Zurich in the 1930s. The Sihl reservoir (Sihlsee) regulates almost half of the Sihl catchment area at the Zurich gauging station. In the city of Zurich the damage potential in the event of extreme flooding of the Sihl, largely exceeding the retention effect of the Sihlsee, has been estimated at up to 6.7 billion CHF. The consequential costs of disruptions in energy supply, telecommunications, and transport would exceed the material damage several times. The Canton of Zurich is therefore implementing a comprehensive plan to improve flood protection around the Limmat and the Sihl rivers as well as for the natural Lake Zurich, whose level can be artificially lowered before an incoming flood. Several immediate measures have already been implemented, including specific structural measures at critical infrastructure, optimisation of emergency planning and emergency organisation, improvements to flood forecasting including anticipatory regulation of Lake Zurich and of the hydropower reservoir Sihlsee, the largest in Switzerland by area, where an overflow of 1 m on the spillway means an additional water retention of about 10 $Mm^3$. In addition, since 2017, a driftwood retention structure in the Sihl river prevents blockages caused by large floating debris at critical points downstream, including the culverts under the Zurich main train station. As a long-term solution to protect the lower Sihl valley and the city of Zurich from extreme flooding of the Sihl, a flood diversion tunnel is under construction to partly transfer flood peaks from the Sihl river (between the Sihlsee and the city of Zurich) into Lake Zurich (AWEL, 2017; FOEN, 2020).

## 6 OUTLOOK ON DAMS AND RESERVOIRS IN SWITZERLAND

### 6.1 *Dams and reservoirs to ensure security and flexibility of electric energy supply*

Storage hydropower from the Swiss Alps contributes significantly to the stability of the national and European electricity grids and is key to the envisaged energy transition to renewables. Hydropower plays a central role in the Swiss energy strategy, especially for the security of the electricity supply in the seasonal balancing. Both hydropower production and storage capacity are to be increased in the coming years and decades. This can generally be achieved through new constructions, expansions and extensions as well as renewal and rehabilitation of existing installations. Since the economically feasible hydropower potential in Switzerland is already exploited to a high degree, the focus is on energy storage in large hydropower reservoirs and a further flexibilization between electricity generation and consumption by means of pumped storage power plants.

By 2040, at least 2 TWh of additional electricity storage capacity should be available from hydropower. To this end, a participatory process has been initiated by the Federal Department of the Environment, Transport, Energy and Communications (DETEC), the so-called "Hydropower Round Table". It was aimed at developing a common understanding between project developers, responsible authorities and other stakeholders of the challenges facing hydropower in the context of the Energy Strategy 2050, the net-zero climate target, security of supply and the preservation of biodiversity. In this process, 33 potential projects, for which the Swiss Federal Office of Energy (SFOE) had obtained brief descriptions and key figures from the project developers, were screened regarding biodiversity, landscape and energy. The projects with the lowest expected specific impacts on biodiversity and landscape per GWh of additional annual storage capacity received the highest scores. To achieve the target of +2 TWh/winter of additional electricity production, 15 projects with the highest scores were finally listed (SFOE, 2021).

The list includes 11 reservoir enlargements through moderate heightening of dams (H) and other adaptation measures, two new reservoirs in the glaciated high mountains (N) and two expansions of existing facilities (N). The two new reservoirs and the two extensions are expected to contribute around 1 TWh/winter, like the 11 dam heightening projects. The total expected additional electricity storage of 2 TWh/winter results mainly from additional production shifting from summer to winter and to a lesser extent from additional production. The projects are further described in Fauriel et al. (2023), including a comparison to recent studies by ETH Zurich (Ehrbar et al., 2018, 2019; Felix et al., 2020; Boes et al., 2021). Accordingly, the construction of a few new dams and the heightening of several existing dams are expected until 2040 or 2050. Thanks to load bearing reserves, the heightening of existing Swiss dams by a certain percentage of their initial maximum height (e.g. 10 %) is often technically feasible by raising the dam on or near its crest only, i.e. the top of the structure itself, without major structural adaptations like a change of dam type. The reserves stem from intentional overdesign of many Swiss dams designed after 1943 following the experiences of the bombing of German dams by the Allies in the second World War.

The Swiss Energy Strategy 2050 poses several technological challenges, which can only be met by adoption of interdisciplinary solutions. Among these, solar energy will play a key role, and installations in Alpine contexts are confirmed to be more and more competitive thanks to their increased efficiency and to a balanced production profile over the year. Thus, the installation of photovoltaic panels on dams and on their corresponding reservoirs is a solution to be considered because of the following advantages: (i) use of existing hydropower infrastructure for installation and grid connection, (ii) reduced impact on the landscape, (iii) speed of realization, and (iv) production pattern that matches well in combination with the flexibility of storage hydropower. An overview of photovoltaic installations on dam faces and reservoirs in Switzerland is provided by Maddalena et al. (2022), Rossetti et al. (2023) and Maggetti et al. (2023).

### 6.2 *Multi-purpose dams and reservoirs*

To date, only a few Swiss reservoirs are explicitly managed as multi-purpose facilities and only a few concessions are linked to further water-related services (Palmieri et al., 2023). With climate change, however, the pressure on water resources will increase due to other usage demands (section 4.4). With regard to the irrigation function, it must be taken into account that most of Switzerland's Alpine reservoirs are located too far away from agricultural areas (Kellner & Weingartner, 2018) and are many times smaller compared to natural lakes (e.g. Grimsel Lake compared to Lakes of Brienz and Thun).

Multi-purpose use of reservoirs creates synergies and is expected to increase public acceptance. For some of the new reservoirs from the list of the Round Table Hydropower, other uses besides hydropower are discussed and planned. For example, the Gorner Lake is to fulfil an essential flood protection function for Zermatt and the Mattertal and shall additionally serve the water supply for irrigation, drinking water, snow production, firefighting, etc. (Fauriel et al., 2023).

The most advanced new multipurpose reservoir project in the Swiss Alps is the Trift dam in the Bernese Oberland, currently in the licensing procedure. Where 75 years ago, at the time of the founding of the Swiss Committee on Dams, the tongue of the Trift glacier was located, a proglacial lake has formed since about the turn of the millennium. The storage volume of this natural lake is to be increased to about 85 $Mm^3$ with a new arch dam (Figure 7). From a water management perspective, such a reservoir can partially replace the storage function of the glacier by temporarily storing the runoff and releasing it to downstream users in a delayed manner or at times of increased demand (geo7 AG, 2017; Kellner & Weingartner, 2018).

Figure 7. Trift glacier (Canton Berne) 1948 (left), 2008 with proglacial lake (centre) and visualisation of the planned arch dam with reservoir (right) (photos: Kraftwerke Oberhasli AG).

### 6.3 *Dam and reservoir adaptation measures due to climate change*

As the climate changes, the variability of runoff increases (Annandale et al., 2016; Palmieri et al., 2023). Annual runoff volumes are also affected by this, as dry and wet years tend to increase in severity (section 4.4). The need for dams and the storage volumes of the corresponding reservoirs to balance water supply and demand will thus increase, especially in regions where natural storage such as from glaciers is declining or diminishing. At the same time, there is still contradictory discussion among experts as to whether sediment input into reservoirs will increase or decrease. While sediment availability tends to increase on the one hand, a decreasing sediment transport capacity is to be expected with decreasing summer run-offs on the other hand.

Impacts of climate change with possible adaptation measures are reported in Table 1.

Table 1. Impacts of climate change with possible adaptation measures.

| Impacts | Possible adaptation measures |
|---|---|
| Increase in magnitude of extreme floods | Capacity increase of spillways, if necessary, including consideration of increased risks of clogging with a trend towards increasing volumes of floating debris (SCD, 2017b; Schmocker & Boes, 2018). |
| Increasing need for flood protection for downstream dwellers | Adaptation of reservoir management and/or the outlet devices, if necessary (see examples of Sihlsee and Mattmark in section 5.2). |
| Changes in inflow volumes and water demands in the region of a dam | Adaptation of reservoir management, and conversion into a multi-purpose facility |
| Intensification of sediment input into reservoirs | Effective reservoir sediment management for sustainable operation, heightening of power water intakes and outlets on dams if necessary. |
| Possibly more frequent and more intense gravitational natural disasters in the area of dams, e.g. mass slides into reservoirs and danger of impulse waves (Evers et al., 2018, 2022). | Increased monitoring of the reservoir slopes, investigation of the effects of impulse waves, disposition for short-term partial drawdowns, if necessary. |
| Changed displacement behaviour of concrete dams due to higher ambient temperatures | Predictive assessment by analytical investigations and modelling; if necessary intensified dam monitoring |

### 6.4 *Dam and reservoir adaptation measures due to new ecological requirements*

The European Water Framework Directive and national legislation place new requirements on transverse structures such as weirs and dams in terms of ecology. In Switzerland, the revised Waters Protection Act entered into force in 2011, which imposes various new or stricter requirements on HPPs. With regard to dams used for hydropower, these are, in particular:

- Reestablishment or improvement of fish migration, both upstream and downstream,
- Reestablishment or improvement of bedload passage and/or bedload management,
- Measures to mitigate hydropeaking effects on the watercourse downstream of storage HPPs.

The Swiss cantons have identified the need for action in strategic planning, and the implementation of the measures for so-called "hydropower rehabilitation" is to take place by 2030. Consideration is given to whether the measures are technically and financially feasible, i.e. whether they have a reasonable cost-benefit ratio. The upstream and/or downstream fish passage must be rehabilitated at around 700 facilities (Dönni et al., 2017), whereby not all of them are considered dams. Sediment passage is to be improved at up to 200 HPPs (Bammatter et al., 2015). While these two aspects apply mainly to low-head facilities on the larger valley watercourses (Boes et al., 2017), hydropeaking countermeasures are to be implemented at around 100 storage power plants, mainly in the Swiss Alpine region (Bammatter et al., 2015).

Rehabilitation of existing HPPs will generate high costs for owners or concessionaires. First and foremost, structural adjustments are planned, but operational measures may also be used, for example temporary drawdown of a reservoir water level to increase the passage of sediment and driftwood during a flood. The HPP operators are compensated by the national grid operator (Swissgrid) for the necessary rehabilitation measures. Eligible non-recurring cost elements are, for example, planning and project costs, acquisition costs for land or buildings, construction costs, costs for new control technology, outage costs and revenue losses due to reduced production.

In the recent past, these requirements have already led to the construction or renovation of ancillary facilities such as upstream and downstream fishways (Meyer et al., 2016), to sediment management measures on dams (Schleiss et al., 2016; Boes et al., 2017) and to new compensation basins for hydropeaking (Schweizer et al., 2021). Examples include a fish lift at the Maigrauge dam in the city of Fribourg, the compensation basin of the Oberhasli power plants in Innertkirchen (Figure 8) and sediment replenishment below the Rossens dam on the Sarine river. There, gravel is deposited on the riverbank below the dam and eroded and transported downstream by means of so-called artificial floods, which are generated by discharge through bottom outlets (Figure 9), see also Friedl et al. (2017).

Many efforts will still have to be made to meet the requirements of the revised Swiss Waters Protection Act. From today's perspective, this will go far beyond the target year of 2030. Among others, new solutions will have to be found, which have been the subject of intensive research projects for years. One example is the protection of fish at large HPPs and water intakes with design discharges well above 100 $m^3/s$, for which there is still no generally recognised technological state of the art (Rutschmann et al., 2022). The use of synergies is another approach to restoring or improving the watercourse continuum for both sediment and aquatic organisms at transverse structures (Foldvik et al., 2022).

Figure 8. Innertkirchen hydropeaking compensation basin (Canton Berne) of Kraftwerke Oberhasli KWO (photo: KWO AG).

Figure 9. Artificial flood at the Rossens dam (Canton Fribourg) of Groupe E (photo: Ecohydrology Research Group, ZHAW).

## ACKNOWLEDGEMENTS

The authors cordially thank Dr. David Felix, AquaSed GmbH, for his careful proofreading and valuable inputs.

## REFERENCES

Annandale, G.W., Morris, G.L., Karki, P. (2016). Extending the Life of Reservoirs: Sustainable Sediment Management for Dams and Run-of-River Hydropower. Directions in Development. Washington, DC: World Bank. doi: 10.1596/978-1-4648-0838-8.

AWEL (ed.) (2017). Hochwasserschutz Sihl, Zürichsee und Limmat (Flood protection Sihl, Lake Zurich and Limmat'). Amt für Abfall, Wasser, Energie und Luft (AWEL), Abteilung Wasserbau. Available at: https://www.zh.ch/de/planen-bauen/wasserbau/wasserbauprojekte/hochwasserschutz-sihl-zuerich see-limmat.html (accessed May 29$^{th}$, 2023; in German)

Bammatter, L., Baumgartner, M., Greuter, L., Haertel-Borer, S., Huber Gysi, M., Nitsche, M., Thomas, G. (2015), Renaturierung der Schweizer Gewässer: Die Sanierungspläne der Kantone ab 2015 ('Restoration of Swiss watercourses: rehabilitation plans of the Cantons from 2015'). Federal Office for the Environment FOEN, Water Department, Ittigen, Switzerland.

Baumer, A. (2012). Innalzamento diga Luzzone: uno sguardo su 15 anni d'esercizio (Luzzone dam heightening: glance of 15 years of development). *Wasser Energie Luft* 104 (3):204–208 (in Italian).

Biedermann, R., Pougatsch, H., Darbre, G., Raboud, P.-B., Fux, C., Hagin, B., Sander, B. (1996). Speicherkraftwerke und Hochwasserschutz ('Storage hydropower plants and flood protection'). *Wasser Energie Luft* 88(10): 220–265 (in German and French).

Boes, R., Hohermuth, B., Giardini, D. (eds.), Avellan, F., Boes, R, Burlando, P., Evers, F., Farinotti, D., Felix, D., Hohermuth, B., Manso, P., Münch-Aligné, C., Schmid, M., Stähli, M., Weigt, H. (2021). Swiss Potential for Hydropower Generation and Storage, Synthesis Report, ETH Zurich, https://doi.org/10.3929/ethz-b-000517823.

Boes, R.M., Albayrak, I., Friedl, F., Rachelly, C., Schmocker, L., Vetsch, D., Weitbrecht, V. (2017). Geschiebedurchgängigkeit an Wasserkraftanlagen ('Bedload continuity at hydropower plants'). *Aqua viva* 59(2): 23–27, http://www.aquaviva.ch/aktuell/zeitschrift (in German).

Brunner M., Björnsen Gurung A., Zappa M., Zekollari H., Farinotti D., Stähli M. (2019). Present and Future Water Scarcity in Switzerland: Potential for Alleviation through Reservoirs and Lakes. *Science of The Total Environment*, 666: 1033–1047. DOI: 10.1016/j.scitotenv.2019.02.169.

CH2018 (2018). CH2018 – Climate Scenarios for Switzerland, *Technical Report*, National Centre for Climate Services, Zurich, 271 p.

Dönni, W., Spalinger, L., Knutti, A. (2017). Erhaltung und Förderung der Wanderfische in der Schweiz – Zielarten, Einzugsgebiete, Aufgaben ('Keeping and fostering migratory fish in Switzerland – target species, catchment areas, tasks'). Study on behalf of the Feral Office for the Environment, 53 pages.

Droz, P. (2023). Swiss dam engineering in the world. *Proc. ICOLD European Club Symposium "Role of dams and reservoirs in a successful energy transition"* (Boes, R.M., Droz, P. & Leroy, R., eds.), Taylor & Francis, London.

Ehrbar, D., Schmocker, L., Vetsch, D., and Boes, R. (2018). Hydropower Potential in the Periglacial Environment of Switzerland under Climate Change. *Sustainability* 10.8: 2794. DOI: 10.3390/su10082794.

Ehrbar, D., Schmocker, L., Vetsch, D., Boes, R. 2019. Wasserkraftpotenzial in Gletscherrückzugsgebieten der Schweiz ('Hydropower potential in regions of retreating glaciers in Switzerland'). *Wasser Energie Luft*, 111(4): 205–212 (in German).

Evers, F.M., Schmocker, L., Fuchs, H., Schwegler, B., Fankhauser, A.U., Boes, R.M. (2018). Landslide generated impulse waves: assessment and mitigation of hydraulic hazards. *Proc. 26th ICOLD Congress*, Vienna, Austria, Q.102-R.40: 679–694.

Evers, F., Boes, R.M. (2022). Slide-induced impulse waves in the context of periglacial hydropower development. *Proc. 27th ICOLD Congress*, Marseille, France, Q.106-R.36: 59–69

Fauriel, J., Filliez, J., Felix, D., Boes, R.M. (2023). Additional water and electricity storage in the Swiss Alps: from studies of potential towards implementation. *Proc. ICOLD European Club Symposium "Role of dams and reservoirs in a successful energy transition"* (Boes, R.M., Droz, P. & Leroy, R., eds.), Taylor & Francis, London.

Felix, D., Müller-Hagmann, M., Boes, R. (2020). Ausbaupotential der bestehenden Speicherseen in der Schweiz ('Potential of extending existing storage lakes in Switzerland'). *Wasser Energie Luft*, 112(1): 1–10 (in German).

Felix, D., Ehrbar, D., Fauriel, J., Schmocker, L., Vetsch, D.F., Farinotti, D., Boes, R.M. (2022). Potentials for increasing the water and electricity storage in the Swiss Alps. *Proc. 27th ICOLD Congress*, Marseille, France, Q.107-R.24: 367–387.
FOEN (ed.) (2020). Regulierung Zürichsee ('Regulation of Lake Zurich'), Fact Sheet. Available at: https://www.zh.ch/de/umwelt-tiere/wasser-gewaesser/hochwasserschutz/hochwasserrueckhalt-seeregulierung.html (accessed May 29th, 2023; in German)
FOEN (ed.) (2021). Auswirkungen des Klimawandels auf die Schweizer Gewässer. Hydrologie, Gewässerökologie und Wasserwirtschaft ('Impacts of climate change on Swiss watercourses. Hydrology, stream ecology and water resources managment'). Federal Office for the Environment FOEN, Bern. *Umwelt-Wissen 2101*: 134 p.
Foldvik, A., Silva, A.T., Albayrak, I., Schwarzwälder, K., Boes, R.M., Ruther, N. (2022). Combining Fish Passage and Sediment Bypassing: A Conceptual Solution for Increased Sustainability of Dams and Reservoirs. *Water* 14(12), 1977, https://doi.org/10.3390/w14121977.
Friedl, F.; Battisacco, E.; Vonwiller, L.; Fink, S.; Vetsch, D.F.; Weitbrecht, V.; Franca, M.J.; Scheidegger, C.; Boes, R.; Schleiss, A. (2017): Geschiebeschüttungen und Ufererosion ('Bedload replenishment and bank erosion'). In: *Geschiebe- und Habitatsdynamik. Merkblatt-Sammlung Wasserbau und Ökologie*. Federal Office for the Environment FOEN, Bern: Bulletin 7 (in German, French and Italian).
geo7 AG (2017): Multifunktionsspeicher im Oberhasli ('Multi-purpose reservoirs in the Oberhasli'). Report on behalf ofWasser und Abfall (AWA) des Kantons Bern. Bern, Switzerland.
Hager, W.H. (2023). Outstanding past Swiss dam engineers. *Proc. ICOLD European Club Symposium "Role of dams and reservoirs in a successful energy transition"* (Boes, R.M., Droz, P. & Leroy, R., eds.), Taylor & Francis, London.
Kellner, E., Weingartner, R. (2018). Chancen und Herausforderungen von Mehrzweckspeichern als Anpassung an den Klimawandel ('Chances and challenges of multi-purpose reservoirs as an adaptation to climate change'). *Wasser Energie Luft*, 110(2): 101–107 (in German).
Leroy, R. (2023). Water resources optimisation – a Swiss experience. *Proc. ICOLD European Club Symposium "Role of dams and reservoirs in a successful energy transition"* (Boes, R.M., Droz, P. & Leroy, R., eds.), Taylor & Francis, London.
Maddalena, G., Hohermuth, B., Evers, F.M., Boes, R., Kahl, A., (2022). Photovoltaik und Wasserkraftspeicher in der Schweiz-Synergien und Potenzial. *Wasser, Energie, Luft* 114(3): 153–160 (in German).
Maggetti, D., Maugliani, F., Korell, A., Balestra, A. (2023). Photovoltaic on dams – Engineering challenges *Proc. ICOLD European Club Symposium "Role of dams and reservoirs in a successful energy transition"* (Boes, R.M., Droz, P. & Leroy, R., eds.), Taylor & Francis, London.
Meyer, M., Schweizer, S., Andrey, E., Fankhauser, A., Schläppi, S., Müller, W., Flück, M. (2016). Der Fischlift am Gadmerwasser im Berner Oberland, Schweiz ('The fish lift on the Gadmerwasser in the Bernese Oberland, Switzerland'). *WasserWirtschaft* 106 (2/3):42–48 (in German).
Muelchi, R., Rössler, O., Schwanbeck, J., Weingartner, R., Martius, O. (2021). River runoff in Switzerland in a changing climate – runoff regime changes and their time of emergence. *Hydrology and Earth System Sciences*, 25(6): 3071–3086.
Palmieri, A., Maggetti, D., Balestra, A. (2023). Multipurpose dams – A European perspective. *Proc. ICOLD European Club Symposium "Role of dams and reservoirs in a successful energy transition"* (Boes, R.M., Droz, P. & Leroy, R., eds.), Taylor & Francis, London.
Pougatsch, H., Schleiss, A.J. (2023). Swiss dams: overview of historical development. *Proc. ICOLD European Club Symposium "Role of dams and reservoirs in a successful energy transition"* (Boes, R.M., Droz, P. & Leroy, R., eds.), Taylor & Francis, London.
Rossetti, E., Maggetti, D., Balestra, A. (2023). Dams and photovoltaic plants – The Swiss experience. *Proc. ICOLD European Club Symposium "Role of dams and reservoirs in a successful energy transition"* (Boes, R.M., Droz, P. & Leroy, R., eds.), Taylor & Francis, London.
Rutschmann, P., Kampa, E., Wolter, C., Albayrak, I., David, L., Stoltz, U., Schletterer, M. (eds.) (2022). *Novel Developments for Sustainable Hydropower*, Springer, Cham: 91–98, https://link.springer.com/book/10.1007/978-3-030-99138-8.
Sander, B., Haefliger, P. (2002). Umbau der Stauanlage Mattmark für den Hochwasserschutz: Schlüsselereignisse und -aktionen, welche die Zeit seit dem Hochwasser von 1993 bis heute prägten ('Conversion of the Mattmark dam for flood protection: key events and actions that have marked the period since the flood of 1993 until today'). *TEC 21*, 128 (36):20–26 (in German).
SCD (2017a). Concrete swelling of dams in Switzerland. *Report of the Swiss Committee on Dams*, https://www.swissdams.ch/fr/publications/publications-csb/2017_Concrete%20swelling.pdf, 78 p.
SCD (2017b). Floating debris at reservoir dam spillways. *Report of the Swiss Committee on Dams*, https://www.swissdams.ch/fr/publications/publications-csb/2017_Floating%20debris.pdf, 78 p.

Schenk, T., Feuz, B., 1992. Die Erhöhung der Staumauer Mauvoisin ('The heightening of Mauvoisin dam'). *Wasser Energie Luft*, 84(10): 245–248 (in German).
Schleiss, A. J., Franca, M.J., Juez, C., De Cesare, G. (2016). Reservoir sedimentation. *Journal of Hydraulic Research*, 54:6, 595–614, DOI: 10.1080/00221686.2016.1225320
Schmocker, L.; Boes, R. (2018). Schwemmgut an Hochwasserentlastungsanlagen (HWE) von Talsperren ('Floating debris at reservoir dam spillways'). *Wasser, Energie, Luft* 110(2): 91–98 (in German).
Schwager, M., Askarinejad, A., Friedli, B., Oberender, P. W., Pachoud, A. J., Pfister, L. (2023). Swiss dam safety regulation: Framework, recent changes and future perspectives. *Proc. ICOLD European Club Symposium "Role of dams and reservoirs in a successful energy transition"* (Boes, R.M., Droz, P. & Leroy, R., eds.), Taylor & Francis, London.
Schweizer, S., Lundsgaard-Hansen, L., Meyer, M., Schläppi, S., Berger, B., Baumgartner, J., Greter, R., Büsser, P., Flück, M., Schwendemann, K. (2021). Die Schwall-Sunk-Sanierung der Hasliaare: Erste Erfahrungen nach Inbetriebnahme und ökologische Wirkungskontrolle ('The hydropeaking remediation of the Hasliaare: First experiences after commissioning and ecological impact monitoring'). *Wasser Energie Luft*, 113(1): 1–8 (in German).
Schweizerisches Baublatt (1990). Staumauer des Stausees von Mauvoisin VS wird erhöht ('Dam of the Mauvoisin reservoir VS is raised'). 57/58: 2–4 (in German).
SFOE (2021). Press release Round Table Hydropower. Swiss Federal Office of Energy SFOE, https://www.admin.ch/gov/de/start/dokumentation/medienmitteilungen.msg-id-86432.html (in German)
St. Moritz Energie (2023). Geschichte und Pioniergeist ('History and pioneer spirit'). https://www.stmoritz-energie.ch/en/about-us/portrait/history-pioneer-spirit.html; accessed on 29 May 2023.
Viviroli, D., Weingartner, R. (2004). Hydrologische Bedeutung des Europäischen Alpenraumes ('Hydrological significance of the European Alpine region'). In: Federal Office for the Environment FOEN (ed.): *Hydrologischer Atlas der Schweiz*. ISBN 978-3-9520262-0-5.
Wohnlich, A., Fankhauser, A., Feuz, B. (2023). Dam heightening in Switzerland. Proc. ICOLD European Club Symposium "Role of dams and reservoirs in a successful energy transition" (Boes, R.M., Droz, P. &Leroy, R., eds.), Taylor & Francis, London.
Zekollari, H., Huss, M., Farinotti, D. (2019). Modelling the future evolution of glaciers in the European Alps under the EURO-CORDEX RCM ensemble. *The Cryosphere*, 13(4),1125–1146.

*Role of Dams and Reservoirs in a Successful Energy Transition – Boes, Droz & Leroy (Eds)*


# Swiss dams: Overview of historical development

H. Pougatsch
*Geneva, Switzerland*

A.J. Schleiss
*Ecole polytechnique fédérale de Lausanne (EPFL), Lausanne, Switzerland*

ABSTRACT: With more than 220 large dams in operation, compared to its surface of some 41′000 km$^2$, Switzerland has a very large fleet. They were erected to meet various economic and protection needs. Their main assignments concern the storage of water for later use, mainly hydropower, and the protection of property, particularly against floods. It is from the 19th century with the growth of the population and the industrial development that the marked beginning of the construction of large dams. This paper describes, over time, the various stages of this development, the main period of which is between 1950 and 1970. Guaranteeing the safety of these storage schemes at all times is essential. A concept based on three pillars (structural safety, monitoring and maintenance, emergency plan) was developed. Periodic safety assessments have led to the undertaking of maintenance and rehabilitation works for several storage schemes. In the future, the monitoring and uprating of existing structures will remain an important task. New projects with the purpose to increase storage for the winter critical season are also planned and are partly already integrated into an expansion process, in particular to meet actually needs for a safe and low carbon energy transition. Of course, research and development remain specific objectives to maintain the competences of dam engineering not only in Switzerland but also for the participation to worldwide development of dams and reservoirs.

## 1 INTRODUCTION

The oldest dam structures still in use date from the nineteenth century. During the twentieth century, the country's economic development and ensuing energy needs had an influence on the rate of construction of dams associated with quite remarkable hydroelectric projects. While Switzerland today has many large dams, this is due to the driving force of eminent engineers who played pioneering roles: F.L. Ritter, H. Juillard, F. Meyer- Peter, H. E. Gruner, Alfred Stucky, Henri Gicot, Giovanni Lombardi. The most active period of dam building occurred between 1950 and 1970 (Figure 1). There are more than 220 dams in Switzerland under the jurisdiction of the Confederation of which 87% are designed to produce hydropower. Other uses include the storing of water for irrigation, supply of drinking water, and the production of artificial snow (3%), biotopes and leisure activities (3%), as well as protective structures for controlling flooding events and retaining sediment (7%). Among these dams, 56% are concrete dams (which can be further split into 54% gravity dams, 41% arch dams, and 5% multiple arch and buttress dams), 31% embankment dams (earth or rockfill), and 13% gated weirs. Twenty-five dams have a height greater than 100 m and four of these are taller than 200 m. The most impressive dams are in the Alps. Finally, it is important to note the existence of several hundred dams and weirs of more modest dimensions and of various types and uses.

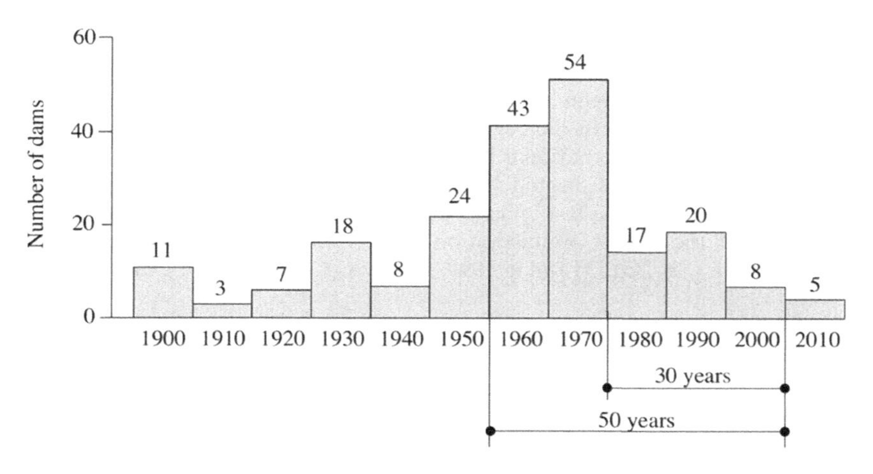


Figure 1. The age structure of large dams in Switzerland.

## 2 MAIN STAGES IN THE CONSTRUCTION OF DAMS

The growth in population and industrial development of the twelfth century led to the appearance of the first water storage structures in Switzerland (Sinniger, 1985). These installations were modestly for the use of hydropower. Only some of these structures still exist today. The industrial revolution in the eighteenth and nineteenth centuries brought about great economic growth, in particular due to the development of hydroelectric schemes in which turbines gradually replaced water wheels. From 1869 to 1872, Guillaume Ritter (1835–1912) built the Pérolles hydropower scheme on the Sarine River upstream of Fribourg. The slightly curved gravity dam (height 21 m, crest length 195 m) was at the time the largest dam in Switzerland. Its trapezoidal section includes two inclined faces. Its construction in concrete was also an innovation in Europe. Interesting rehabilitation works across the whole dam structure were carried out from 2000 to 2004.

In the late nineteenth and early twentieth centuries, many embankment dams were built (Figure 2). At that time, techniques for earthworks were based on empirical criteria far removed from the later science of soil mechanics. The failure of a small, five-meter dike in 1877 encouraged engineer Friedrich de Salis (1828–1901) to undertake a detailed study of the dam break wave, which was most probably the first calculation of its kind. Some of the more remarkable constructions include the Gübsen saddle embankment (SG/1900/H = 19 m) and the Klöntal embankment (GL/1910/H = 30 m), which was designed to raise the water level of a natural lake. During this same period, many gravity dams were also built, most often using traditional masonry techniques, including dams at Buchholz (SG/1892/H = 19 m), List (SG/1908/H = 29 m), Muslen (SG/1908/H = 29 m), and at the Bernina Pass (GR/1911/H = 15 and 26 m). Substantial rehabilitation works were carried out on the dams at the Bernina Pass in the late twentieth and early twenty-first centuries.

Due to increasing energy needs, from 1914 attention turned to the construction of water-storage reservoirs, in particular water issuing from summer snowmelt and glacier melt, and to guarantee the production of energy in winter when demand for electricity rises. This transfer of energy necessitated the availability of large reservoirs and therefore the construction of large dams.

The construction of the Montsalvens dam (FR/1920/H = 55 m) marks the beginning of a new period of dam development. It was the first double-curvature arch dam in Europe and was designed by Heinrich E. Gruner (1873–1947. Static calculations were based on a trial-load method devised by Hugo F. L. Ritter (1883–1956), which was then developed by Alfred Stucky (1892–1969) and Henri Gicot (1897–1982), who were both working with H. E. Gruner.

The Swiss Commission on Dams was founded on 2 October 1928 by six renowned scholars and practitioners in the construction of dams namely H. Eggenberger, H. E. Gruner, A. Kaech, E. Meyer-Peter, M. Ritter, A. Stucky, A. Zwygart and W. Schurter. Somewhat later the renown experts J. Bolomey, O. Frey-Bär, H. Gicot, H. Juillard, M. Lugeon, E. Martz, M. Roš, and M. Roš. jun. joined the commission. According to its statutes, the aim of the commission was to "deal with problems related to dams and to collect information and knowledge about these constructions and their operation." In the beginning, the Commission dealt with issues brought up in conferences held by the International Commission on Large Dams (ICOLD), founded in 1928. It later went on to establish guidelines for the construction and maintenance of Swiss dams. In addition, it set itself the task of carefully examining the results of observations carried out on dams. In 1948 the Swiss Commission on Large Dams became the Swiss National Committee on Large Dams (SNCOLD) and in 1999, the Swiss Committee on Dams (SwissCoD).

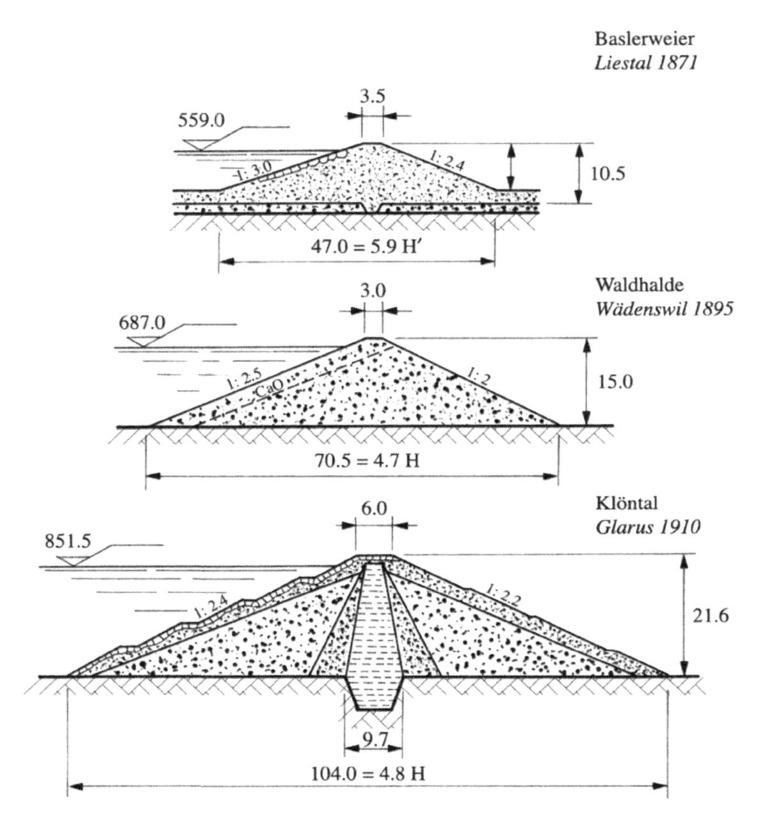


Figure 2. Examples of embankment dams built in the late nineteenth and early twentieth centuries (from Schnitter, 1985).

In the 1930s, the global economic crisis slowed the rise in the consumption of electricity, as well as the need for new hydropower schemes, and, as a result, the construction of dams. However, it was at this time that the Dixence dam was built (VS/1935/H = 87 m), based on a project by Alfred Stucky. This buttress dam has a volume of 421'000 m$^3$ and until the end of the Second World War was the highest dam of its type in the world. Construction of the Verbois dam (GE/1943/H = 34 m) and the buttress dam at Lucendro (TI/1947/H = 73 m) began during the Second World War.

After the bombing of German gravity dams by the English air force in May 1943, the Federal authorities stipulated that horizontal buttress struts had to be added to the already completed Lucendro buttress dam, as well as major reinforcement to the Cleuson dam (VS/H = 87 m) built between 1947 and 1950. Thereafter, only very narrow hollows were to be admitted, as is the case for the dams at Räterichsboden (BE/1950/H = 94 m) and Oberaar (BE/1953/H = 100 m).

During the Second World War, a sharp increase in the consumption of electricity was observed, which continued in the post-war context. Hydroelectric schemes, of which large dams were the centerpiece, were constructed to respond to this demand for additional energy. In 1945 the construction of the Rossens arch dam began (FR/1947/H = 83 m). From 1950 and throughout the 1970s, dam construction boomed, and more than one hundred dams were commissioned. The construction of the 237-meter-high Mauvoisin arch dam (Figure 3) heightened to 250 m in 1991 and the 285-meter-high Grande Dixence gravity dam (Figure 4) began in 1951. Commissioned in 1957 and 1961 respectively, these two dams are still today among the highest operating dams in the world. Thanks to the skills of renowned experts and engineers from high-level consultancy firms, many impressive arch dams were erected.

Figure 3. The Mauvoisin arch dam (H = 250 m).

Figure 4. The Grande Dixence gravity dam (H =285 m).

In the design of arch dams, the traditional circular arches were replaced by parabolic or elliptic arches in order to ensure a better orientation of the pressure of the arches against the rock foundation and abutments. Among the structures with a height greater than 100 m, the following dams can be listed: Emosson (VS/1974/H = 180 m), Zeuzier (VS/1957/H = 156 m), Curnera (GR/1966/H = 155 m), Zervreilla (GR/1957/H = 151 m), Moiry (VS/1958/H = 148 m), Limmern (GR/1963/H = 146 m), Punt dal Gall (GR/1968/H = 130 m), Nalps (GR/1962/H = 127 m), and Gebidem (VS/1967/H = 122 m). The twin-arch dams at Lake Hongrin (VD/1969/H = 125 and 90 m) should also be noted, whose double arches joined by a common, central abutment, give the structure a distinctive look (Figure 5). After the Grande Dixence and Mauvoisin dams, the 200 m barrier was broken, notably by the arch dams of Luzzone (TI/1963/1997/H = 225 m) and Contra (TI/1965/H = 220 m) (Figure 6) in Ticino. The latter was designed by Giovanni Lombardi (1926–2017).

Utilizing primarily the core scientific and rational developments established by Karl Terzaghi, the founder of soil mechanics, the construction of several embankment dams was

Figure 5. The double-arch Hongrin dam (H = 125/90 m).

undertaken. In addition, on the initiative of Eugen Meyer-Peter (1883–1969), the ETHZ created an institute for foundation engineering and soil mechanics, whose research has proven to be of great use. The embankments at Marmorera (Castiletto), (GR/1954/H = 91 m), Göscheneralp (UR/1960/H = 155 m), and Mattmark (VS/1967/H = 120 m) are the largest embankment dams built in Switzerland (Figure 7).

Figure 6. The Contra arch dam (H = 220 m).

From 1980, new constructions integrated into hydroelectric schemes became less common. The largest are the structures in Solis (GR/1986/H = 61 m) and Pigniu (GR/1989/H = 53 m). However, to better operational conditions and ensure improved energy transfer, the arch dam at Mauvoisin (VS) was raised by 13.5 m to reach a height of 250 m in 1991 and the dam at Luzzone (TI) by 17 m to reach 225 m in 1997. Two pumped-storage projects have been launched. The first is the scheme at Limmern (GL); here, the construction of the Muttsee gravity dam (GL /2015/ H = 35) has raised the level of a natural lake, situated at an altitude of 2,500 meters above sea level, by 28 m. During construction of the Nant-de-Drance pumped-storage hydropower plant (VS), the Vieux-Emosson Dam, built in 1955, was raised by 21.5 m to reach a height of 76.5 m, which has doubled the storage capacity of the reservoir.

Furthermore, several structures with heights between 7 and 30 m have been built to protect against natural events such as floods and avalanches. In the late 1990s, reservoirs created to store water with the aim of producing artificial snow began to appear.

Figure 7. The largest embankment dams in Switzerland: (a) Mattmark (H = 120 m), (b) Göscheneralp (H = 155 m), and (c) Marmorera (Castiletto) (H = 91 m).

## 3 REINFORCEMENT AND REHABILITATION WORKS

On a different note, from the early 1980s, attention turned toward old dam structures of all sizes whose safety had to be reassessed on the basis of modern standards and technological advances. Depending on the results obtained, it may be necessary to rehabilitate all or some of the structure elements in order to guarantee its safe operation for many more years. Of course, there are many reasons why the strengthening and rehabilitation of a dam are necessary. Often, the structure no longer meets the latest stability criteria. The accepted hypotheses for load on the structure from the initial project may have to be revised. These hypotheses may concern the weight itself, the distribution of uplift or the induced effects that may occur during an earthquake. New operational conditions, such as, for example, new flooding levels, a large accumulation of upstream sediment, or the installation of downstream rockfill may also have an impact on load. Several different types of rehabilitation works are possible and may sometimes be combined. Possible interventions reinforcement and rehabilitation interventions are:

- Dam heightening
- Complete rehabilitation of all structures
- Treatment of facing
  - Laying membrane
  - Asphalt facing
  - Concrete cover, gunite
- Rehabilitation of dam body (concrete, embankments)
  - Grouting
  - Sealing

- Reinforcement of downstream toe
- Foundation treatment
  - Grouting
  - Drainage
- Flood safety
  - Modification of the spillway
  - Modification of the crest
  - Creation of a parapet wall
- Drawdown of the reservoir
  - Transformation of the bottom outlet
  - Implementation of a new bottom outlet

In the case of the Maigrauge dam (FR/H = 22 m), commissioned in 1872, the rehabilitation project was designed to improve safety in case of flooding by uprating the spillway, improving the dam's structural safety with the installation of prestressed rock anchors, and optimizing operating conditions with the modification of water intakes. Furthermore, the monitoring system was modernized. And last but not least, a ladder including a lift was constructed for fish to migrate upstream past the dam and a series of channels and pools were added for downstream migration. These works were carried out from 2000 to 2004. When the stress and stability assessment demonstrate that safety conditions are not being met, drawing down the reservoir or reinforcing the dam becomes inevitable. For the Gübsensee dam (SG/1900/H = 24/17 m), the chosen solution was to install post-stressed cables. As for the concrete dams at Muslen (SG/1908/H = 29 m) and List (SG/1908/H = 29 m), their upstream and downstream faces were covered with a concrete shell while the height of the crest was also raised in order to increase the volume of the reservoir and thus optimize hydropower generation.

Remedial work may become necessary if the material at the core of the dam has suffered from major internal damage, due, for example, to swelling caused by an alkali-aggregate reaction (AAR), which can significantly impact concrete characteristics. To limit the development of this type of swelling, the upstream face of the Illsee dam (VS/1923–43/H = 25 m) was lined with a PCV geomembrane, and the Lago Bianco Sud dam (GR/1912–42/H = 26 m) was lined with a membrane comprising a synthetic liquid applied in successive layers. At the Illsee dam, the system put in place did not prevent water penetration via the foundation and thus did not slow the swelling phenomenon. Concrete drying measures were also taken without success. Vertical sawing cuts into the concrete are still to be done in order to relieve the stresses in the dam. After being commissioned in 1952, monitoring of the behavior of the Serra arch dam (VS/1952–2010/H = 25.7 m) was principally carried out through geodetic measurements and leveling. Concrete in the Serra dam, affected by an alkali-granulate reaction (AAR) leading to the swelling of the concrete, resulted in irreversible upstream deformation, accompanied by an uplift to the dam and diffuse cracking. Rehabilitation of the structure was necessary, as the gradual deterioration in its conditions of use and safety had been highlighted. The rehabilitation solution that was chosen consisted in building a new dam downstream of the original one (Figure 8). The partial and necessary demolition of the downstream toe of the old dam enabled a new and more favorable geometry to be determined. From a structural point of view, the Serra dam is close to a double-curved arch dam (SwissCoD, 2017b).

In the 1930s, the designers and builders of the Spitallamm dam (BE/1932/H = 114 m) did pioneering work. The dam near the Grimsel pass is one of the first large arch-gravity dams. The Seeuferegg gravity dam (BE/1932/H = 42 m) was built at the same time as the Spitallamm dam. These two dam walls created the reservoir of Lake Grimsel. In the 1960s, detailed examinations and checks revealed that the Spitallamm dam had a vertical crack in its core. In fact, the crown and the concrete of the downstream face of the Spitallamm dam had started to separate from the rest of the dam. Initially, it was decided to carry out the necessary rehabilitation work as part of the possible raising of the two Grimsel dams. Due to a possible alkali-aggregate reaction, the operator decided not to rehabilitate the dam and, in the fall of 2015, started the planning work for a new dam. In June 2019, the construction of a new double-

Figure 8. Serra Dam during reconstruction: new arch dam in front of the old dam before its partial demolishing.

Figure 9. Spitallamm dam under construction in front of the existing arch-gravity dam (Courtesy A. Schleiss, 2022).

curved arch dam located immediately in front of the old dam began. The old Spitallamm dam will be left as it is and will subsequently be submerged (Figure 9).

In cases of significant deterioration due mainly to frost, both faces must be treated. For example, around 1983, after damaged zones had been scored, the surface of the upstream face of the Schräh dam was covered with lightly reinforced shotcrete to a depth of 8 to 12 cm. In another example, wet shotcrete was applied across practically the whole surface of the upstream face of the Cleuson dam between 1995 and 1998. The affected area had previously been stripped by hydro-demolition. The bond between the base concrete and the sprayed concrete was guaranteed by a grid with mushroom-shaped anchoring bolts made of 12-mm-diameter reinforced steel placed at 4 bolts per $m^2$.

After severe flooding in 1978 in Ticino and the blockage of the spillway channels at the Palagnedra dam (TI/1952/H = 72 m) by the massive piling up of driftwood (Figure 10) (SwissCoD, 2017a), the supervisory authority reviewed general safety criteria in cases of flood and asked for an investigation into the safety requirements to be carried out. It is important to note that in the Alps, floods can strike with devastating speed. For many dams, spillways or crests had to be modified, or a large parapet wall had to be added so as to provide increased retention capacity.

Studies have shown that during major flooding events, reservoirs attached to hydroelectric schemes contribute substantially to the reduction of peak flooding levels, due to their capacity for retaining water, even though that is not their primary function. In order to increase

Figure 10. Blocked passes at the Palagnedra dam (TI) during the flood on August 7, 1978.

protective measures against flooding events downstream of a dam, without having to restrict production of hydroelectric power, one option is to transform a single operation into a multipurpose operation. The idea is to create an additional volume in the upper part of the reservoir that can be used hold a specific volume of water in cases of flood. Such a project was completed in 2001 for the Mattmark embankment dam by adapting the side weir spillway. Other proposals are also being considered.

In Switzerland, the general rule is that all dams must be equipped with a bottom outlet to empty the reservoir in cases of abnormal dam behavior or lower the water level for maintenance to be carried out. Due to insufficient capacity or obsolete equipment, some structures had to be completely transformed or a new bottom outlet had to be created. For example, a gallery was drilled into the foot of the Schräh dam (SZ/1924/H = 111 m). To ensure compliance, a bottom outlet was created in the Illsee dam (VS/1924–43/H = 25 m) by utilizing a gallery that had originally been used to lower the level of the natural lake at the time the dam was being built. In other cases, gates have simply been replaced or a new gate added.

Geodetic measurements carried out between 1921 and 1937 had already highlighted weak plastic movement perpendicular to the bed of the downstream section of the left-bank abutment of the Montsalvens dam (FR/1920/H = 52 m). As this movement continued, additional monitoring devices (pendulum, extensometers) were installed in 1969 in order to ensure more systematic reporting of the behavior of the downstream zone. Analyses carried out later showed that the state of equilibrium was situated at the limit of elastic behavior. Strengthening works were decided on out of a fear that the deformations would only continue to intensify or that a rockfall would be caused by a seismic shock. Works on the left abutment were designed with two objectives in mind; firstly, bolts sealed with cement grout were applied to a concrete sprayed surface to protect the valley flanks against the risk of rock fall, and secondly, reinforced bars sealed completely into the rock with cement grout were designed to increase resistance to shearing along bedding planes. At the Pfaffensprung dam (UR/1921/H = 32 m), it was not known how long existing anchors would hold downstream of the left bank abutment, so it was decided that additional prestressed anchors would be installed with a system that enabled their tension to be controlled at all times.

## 4 DEALING WITH RESERVOIR ISSUES

### 4.1 *Reservoir surroundings*

It is of the utmost importance to monitor the behavior of banks and slopes, as instabilities can occur, sometimes without any direct connection to activity at the reservoir. For example, upstream of the Mauvoisin dam, a crack was observed in a mountain road running alongside the reservoir. Snowmelt had saturated scree in the area, which had slipped in sections of differing depths. A monitoring system (geodetic measurements, inclinometric measurements from boreholes) was set up, and a limited water level was set while the zone was still unstable. Similarly, experts inspected several glaciers in order to ensure that large sections of material did not break off and end up in the reservoir. With progressing climate change, it has to be expected that the risk of instabilities of reservoir bank and slopes will increase.

### 4.2 *Sedimentation*

Due to climate change and its consequences (glacier retreat, the zero-degree line and permafrost levels rising in altitude, increased precipitation), an increase in the arrival of solid materials into alpine lakes must be expected. Solid materials issuing from soil erosion are transported toward water reservoirs in watercourses by bed load or suspended load. Whether this material settles in a particular place or spreads out into the reservoir depends on its size. This deposited sediment has a direct impact on the operation of the overall storage scheme, as well as on the safety of the dam. With regard to operation, this will above all be manifested in a loss of usable storage capacity. According to estimations, at a global level this reservoir volume loss is between 1 and 2% per year. Based on an analysis of 19 reservoirs (Beyer, Portner, and Schleiss, 2000), it is in the order of 0.2% for alpine storage schemes in Switzerland. The siltation of reservoirs can affect the lifetime of a dam. With regard to safety issues, there is a considerable risk of water intakes and bottom outlets in particular being obstructed. These situations must be avoided. For bottom outlets to remain operational at all times, a free space directly upstream of the discharge system must be guaranteed. For various reasons, it is vital that an appropriate amount of this deposited sediment be periodically removed. Several means for limiting the arrival of sediment into reservoirs already exist (settling basins, diversion galleries, sand traps, etc.). Sediment by-pass tunnels are successfully in operation at several dams (Palagnedra, Pfaffensprung, Rempen, Runcahez, Solis) with more that 100 years of experience at Pfaffensprung dam (Boes, 2015). In many cases (Gebidem, Rempen, Palagnedra, Luzzone, etc.), a program of periodic flushing takes place in accordance with a predefined schedule. The legal basis regarding the protection of water establishes the terms relative to the flushing and emptying of reservoirs.

Specifically, it is important to ensure that as far as is possible flora and fauna in the river downstream are not harmed during these operations. Furthermore, except in extraordinary events, permits are issued by the relevant cantonal authorities, some of which have established regulatory requirements. In the future, designers and operators will be required to take effective measures for preventing reservoir sedimentation. In alpine reservoirs, turbidity currents that form during floods are responsible for the transportation of considerable amounts of fine particles of sediment along the reservoir. Turbidity currents, which are like underwater avalanches, also erode sediment that has already been deposited, bringing it nearer to the dam itself where it is more likely to block the entrance to bottom outlets or water intakes (Schleiss and Oehy, 2002; Oehy and Schleiss, 2003). The increase in reservoir sedimentation can force operators to undertake substantial work in order for these structures to retain their primary functions. For example, at the Mauvoisin dam, the water intake had to be raised by 38 m and the bottom outlet by 36 m by building new intakes and gate chambers. Many other cases also exist.

Fine sediment, principally transported along the bottom of the reservoir by turbidity currents, can contribute to more than 80% of sedimentation in alpine reservoirs. In addition to controlling turbidity currents in reservoirs through the use of obstacles (Oehy and Schleiss, 2003), the deposit of fine material in the vicinity of the dam can be avoided by venting it through bottom outlets. This approach is economically and environmentally beneficial. Artificial flood releases combined with sediment replenishment downstream of dams can work together with this venting of fine sediment and restore, as well as dynamize, bedload transport in the river downstream (Döring et al., 2018). Another promising option for the management of fine sediment is by using water jet installations in the reservoir near the dam to ensure its suspension before evacuating it at controlled concentrations through the powerhouse intake (Jenzer, Althaus et al., 2011).

## 5 BRIEF REMINDER OF THE LEGAL BASES

As regards the safety of dams, the Swiss supervisory authority has two goals. Firstly, that of ensuring the safety of the dam and therefore that of the population, and secondly, of ensuring the safety of the operation itself. From a historical point of view, the June 22, 1877 Water Regulation Act amended in 1950 outlines the provisions relating to dam safety. It stipulates in one of its primary articles that concerning dams the Federal Council must take all necessary measures to prevent, insofar as is possible, any danger and damage that may result from their means of construction, their inadequate maintenance, or from acts of war. Currently, the legislative basis includes an act on dams (WRFA, 2010) in effect since 2013, accompanied by an ordinance (WRFO, 2012). In addition to the terms relating to their safety, the act introduces the notion of civil liability due to the risks.

In terms of scope, the act (WRFA, 2010) specifies that the liability applies:

- To dams whose water level H above the low-water level of the watercourse or the thalweg (reservoir height) is at least 10 m, or
- If this water level is at least 5 m, for those whose reservoir capacity is higher than 50,000 $m^3$
- To dams of smaller dimensions, if they represent a specific potential risk for people and property; otherwise they are exempt.

## 6 MEASURES TO GUARANTEE PUBLIC SAFETY

Following the bombing of German gravity dams located in the Ruhr valley by the English air force during the nights of May 16 and 17, 1943, Swiss military and civil authorities were concerned about the vulnerability of dams to acts of war or sabotage. An initial measure introduced in June 1943 was to suspend a cable above dams as a means of protection against airplanes. In September 1943 the Federal Council, who had worked quickly on the legislation, announced an ordinance whose provisions covered active and passive dam protection measures against destruction in times of war, the use of reservoirs, and the lowering of their level, as well as the installation of an alarm system. It was decided that sirens would initially be installed in the near zone—the area subject to flooding 20 minutes after destruction of the dam. A list of dams that had to be equipped with the alarm system was published in late November 1943.

In 1945 the Bannalp (NW/1937/H = 32 m) and Klöntal (GL/1910/H = 30 m) embankment dams were the first to be fitted with the flood wave alert system. The regulation concerning dams that came into effect in July 1957 gave a legal basis to the recommendation that alarm systems should be installed. However, this system still needed to be improved, and a technical committee was mandated to establish the terms of reference for a new system. A more concrete definition of the flood wave alert system was introduced in the 1957 version of the regulation concerning dams. For the first time, the near zone identified was extended to 2 hours and a far zone was designated. The alarm systems for each zone are different. In addition, it was decided that the flood alert system would also be used in peace time and extended to all other possible hazards to dam safety. The introduction of degrees of preparation and the definition of criteria for the triggering of the flood alert system were also new elements.

## 7 LOOKING TO THE FUTURE

The saga of the construction of large dams in Switzerland is now practically over, as the technically most interesting sites are mostly exploited. The most recent major dam project in Switzerland was the heightening between 1995 and 1997 of the Luzzone arch dam, built in the early 1960s and whose height was increased from 208 to 225 m. Another large-scale project concerning the raising of the Grimsel reservoir level by 23 m (Spitallamm and Seeuferegg dams) is under study and should come to fruition.

In order to support the energy transition defined by the 2050 energy strategy, the Swiss government organized a round table in 2021, including civil society, to define projects with the objective of ensuring the security of energy supply in winter while preserving biodiversity and the landscape. With the aim of increasing the flexible production of storage facilities in winter by at least 2000 GWh until 2040, the participants agreed on a list of 15 priority projects. This list includes three new dams in valleys freed up by the retreat of the glaciers (Figure 11), e.g. the future Trift dam, and the rest are extensions of existing facilities, including dam heightening.

Figure 11. Photomontage of the future 180 m high Trift dam, which uses the already freed valley by glacier retreat. Above: with empty lake with glacier position today. Below: with full lake (Curtesy KWO).

These last examples show that the field of dams remains attractive for designers, builders and operators. Indeed, by the diversity of its assignments, it offers interesting perspectives not only for projects of transformation or modernization of existing works, but also for new projects. It is also a question of ensuring the good health of the existing works.

With the aging of structures and equipment, some problems must be examined, such as the evolution of mass concretes (long-term creep, alkali-aggregate reaction) or the behavior of foundations (development of under-pressures, preservation of networks of drainage).

In the short term, the monitoring and maintenance of dams of all sizes remain essential tasks in order to guarantee their safety. While the organization of these activities for large and medium structures has been in place for many decades, that of small structures still needs to be regulated. With regard to hydroelectric developments, the future of hydraulic power is part of the framework of sustainable development. Several socio-economic and ecological parameters will influence operators in their future investments, namely the opening of the market, the evolution of supply and demand, electricity prices and construction costs, hydraulic royalties and politics.

First of all, many concessions are coming to an end. When renewing them, assessing the safety of structures that are sometimes dilapidated is a necessary phase which in most cases leads to having to consider major reinforcement and rehabilitation work, or even the replacement of the electromechanical equipment of the landfill structures. Other measures can be taken before this deadline to proceed with the modernization and optimization of existing installations. The idea of raising a dam to increase the capacity of its accumulation basin is a realistic option. Pumped storage projects are back in the news. They have the advantages of storing hydraulic energy and recovering the basic energy of non-adjustable power stations (thermal, solar, wind) produced outside consumption hours. Some thirty potential sites were assessed during the 1970s. Today, efforts are being made to combine this type of structure with existing storage facilities by equipping them with new water supply systems and increasing the retention capacity. For example, the Nant de Drance pumped storage project, which operation started in 2022, uses the difference in level between the Emosson (VS / 1974/H = 180 m) and Vieux Emosson (VS / 1974/2017/H = 76.5 m) reservoirs.

Although the majority of reservoirs were created with a view to producing hydraulic energy, the construction of water storage basins for the production of artificial snow developed strongly towards the end of the 1990s. These structures are generally located outside the river, on a flat area or on the side of a hill; much of the artificial basin can be made by excavation. Given the local conditions and the availability of materials, the use of an embankment dam is frequent. There is no doubt that ski lift operators will still continue to use this means to guarantee the snow cover of the slopes as well as possible. Finally, hydraulic development can also promote the creation of biotopes and recreational areas.

Every year natural disasters (floods, avalanches, debris flows) cause substantial damage and result in considerable costs. The increased need for safety will lead to the design and construction of new protection systems including flood retention works and anti-avalanche dikes. Some protective measures will be revisited, improved, or completed. In Switzerland, reservoirs for existing hydroelectric schemes are generally speaking single-purpose reservoirs. They can be converted into multipurpose reservoirs by setting aside a clearly specified storage volume for retaining water during floods. This solution has already been implemented in certain cases and others may well follow. Various solutions are possible for maintaining the available capacity of reservoirs when necessary, including dam heightening, creating a supplementary connected reservoir, or by adding a seasonal pumped-storage system. It is possible to refer to a flood forecasting model so as to better manage reservoirs. The canton of Valais has taken this approach and developed the Minerve project, which simulates the overall hydraulic behavior of catchment areas and hydropower schemes in Valais. The model is designed to help cantonal officials make decisions (Garcia Hernandez et al., 2011, 2014; Jordan et al., 2008; Raboud et al., 2001).

Ongoing research and development over many years has led to technological advances. Various problems have been tackled such as safety under dynamic loading during earthquakes, extreme flood events, the long-term behavior of dams, and the behavior of foundations. Of course, these issues have not yet been fully resolved. Problems related to safety and the overall behavior of dams continue to occupy researchers in such domains as the behavior of overtopped dams, the long-term behavior of facing and drainage, and reservoir sedimentation. The development of data collection and methods for analysis and data measurement continue to be used in the monitoring of dams. And finally, new and increasingly applied construction methods, such as roller compacted concrete (RCC) and cemented soils, are the focus of specific studies. Swiss expertise in the field of dams is recognized internationally and can thus be employed elsewhere in the world. Global demand for the construction of hydraulic schemes and dams in particular is high and will continue to grow. Logically, the strategy which should be implemented is one that looks outward (Schleiss, 1999). Swiss industry and engineering are capable of achieving this vision thanks to more than one hundred years of experience in hydraulic construction and the international renown garnered since the 1960s by the construction of over 180 large dams outside of Switzerland.

## ABBREVIATIONS OF SWISS CANTONS

ZH Zurich, GL Glarus, AR Appenzell Outer-Rhodes, VD Vaud, BE Bern, ZG Zug, AI Appenzell Inner-Rhodes, VS Valais, LU Lucerne, FR Fribourg, SG St. Gallen, NE Neuchâtel, UR Uri, SO Solothurn, GR Grisons, GE Geneva, SZ Schwyz, BS Basel, AG Aargau, JU Jura, OW Obwalden, BL Basel Disctrict, TG Thurgau, NW Nidwalden, SH Schaffhausen, TI Ticino.

## ACKNOWLEDGEMENT

This contribution is essentially based on Chapter 2.2 and 2.3.2 of the book *Design, Safety and Operation of Dams* by Schleiss A. et Pougatsch H., 2022, EPFL Press, Presse polytechnique et universitaire romande. https://www.epflpress.org/produit/1407/9782889154852/design-safety-and-operation-of-dams.

Most of the dam pictures have been published in the calendars "Dams in Switzerland" issued yearly since 2005 by the Swiss Committee on Dams.

## REFERENCES

Beyer Portner N., Schleiss A., 2000. Bodenerosion in alpinen Einzugsgebieten in der Schweiz. Wasserwirtschaft 90(2),88–92.

R. M. Boes (ed)., 2015. Proc. 1st Int. Workshop on Sediment Bypass Tunnels Zurich. VAW Mitteilung 232, Versuchsanstalt für Wasserbau, Hydrologie und Glaziologie, ETH Zurich, Zürich, 258 pages.

Döring M., Tonolla D., Robinson CH. T., Schleiss A., Stähly S., Gufler CH., Geilhausen M. et Di Cugno N., 2018. Künstliches Hochwasser an der Saane – Eine Massnahme zum nachhaltigen Auenmanagement. wasser, energie, luft – eau, énergie, air, 110. Jahrgang, Heft 2, 119–127.

García Hernández J., Schleiss A. J. et Boillat J.-L., 2011. Decision Support System for the hydropower plants management: the MINERVE project. Dams and Reservoirs under Changing Challenges – Schleiss & Boes (Eds), Taylor & Francis Group, London, 459–468. ISBN 978-0-415-68267-1.

García Hernández J., Claude A., Paredes Arquiola J., Roquier B. et Boillat J.-L., 2014. Integrated flood forecasting and management system in a complex catchment area in the Alps – Implementation of the MINERVE project in the Canton of Valais. Swiss Competences in River Engineering and Restoration, Schleiss, Speerli & Pfammatter Eds, 87–97. Taylor & Francis Group, London, ISBN 978-1-138-02676-6, doi:10.1201/b17134-12.

Jenzer Althaus J., De Cesare G. et Schleiss A. J., 2011. Entlandung von Stauseen über Triebwasserfassungen durch Aufwirbeln der Feinsedimente mit Wasserstrahlen. wasser, energie, luft – eau, énergie, air, 103. Jahrgang, Heft 2, 105–112.

Jordan F., García Hernández J., Dubois J. et Boillat J.-L., 2008. Minerve – Modélisation des intempéries de nature extrême du Rhône valaisan et de leurs effets. Communication n° 38 du Laboratoire de constructions hydrauliques, LCH-EPFL, Ed. A. Schleiss, Lausanne.

Oehy Ch. et Schleiss A., 2003. Beherrschung von Trübeströmen in Stauseen mit Hindernissen, Gitter, Wasserstrahl- und Luftblasenschleier. wasser, energie, luft – eau, énergie, air, 95. Jahrgang, Heft 5/6, 143–152.

Raboud P.-B., Dubois J., Boillat J.-L., Costa S. et Pitteloud P.-Y., 2001. Projet Minerve – Modélisation de la contribution des bassins d'accumulation lors des crues en Valais. wasser, energie, luft – eau, énergie, air, 93e année, cahiers 11/12, 313–317.

SwissCoD, 2017a. Swiss Committee on Dams, Floating debris at reservoir dam spillways. Report of the Swiss Committee on Dams on the state of floating debris issues at dam spillways. Working group on floating debris at dam spillways, November 2017.

SwissCoD, 2017b. Swiss Committee on Dams, Concrete swelling of dams in Switzerland. Report of the Swiss Committee on Dams on the state of concrete swelling in Swiss Dams. AAR Working Group. May 2017.

Schleiss A. et Oehy Ch., 2002. Verlandung von Stauseen und Nachhaltigkeit. wasser, energie, luft – eau, énergie, air, 94. Jahrgang, Heft 7/8, 227–234.

Schleiss A., 1999. Constructions hydrauliques. Facteur clé de la prospérité économique et du développement durable au XXIe siècle, IAS n° 11, 9 juin 1999, 198–205.

Schnitter, N.J., 1985. Le développement de la technique des barrages en Suisse. Comité national suisse des grands barrages, Barrages suisses – Surveillance et entretien, publié à l'occasion du 15e Congrès international des grands barrages, Lausanne 1985, 11–23.

Sinniger, R., 1985. L 'histoire des barrages. EP FL, Polyrama, 1985, 2–5.

WRFA, 2010. Water Retaining Federal Act (in French, German, Italian). RS 721.101 of 1st October 2010 (State in 1st January 2013).

WRFO, 1998. Water retaining Federal Ordinance (in French, German, Italian). December 7, 1998.

WRFO, 2012. Water retaining Federal Ordinance (in French, German, Italian). RS 721.101.1 of October 17, 2012 (state in 1st April 2018).

*Role of Dams and Reservoirs in a Successful Energy Transition – Boes, Droz & Leroy (Eds)*
*© 2023 The Author(s), ISBN 978-1-032-57668-8*

# Outstanding past Swiss dam engineers

Willi H. Hager
*VAW, ETH Zurich, Zürich, Switzerland*

ABSTRACT: Switzerland is a country rich in waterpower, yet without hardly any other resources. The electricity production, therefore, was based on hydropower a long time ago. To store energy, dams are key structures for hydropower schemes. For both low and high head dams these were provided by keen engineers who oversaw the activities relating to dam construction. This paper highlights five selected engineers having significantly contributed to Swiss dam engineering. Of relevance is of course their role within the Swiss National Committee on Large Dams.

RÉSUMÉ: La Suisse est un pays riche en énergie hydraulique, mais sans pratiquement aucune autre ressource. La production d'électricité était donc basée sur l'énergie hydraulique il y a longtemps. Pour stocker l'énergie, les barrages sont des structures clés pour les projets hydroélectriques. Pour les barrages à basse et haute chute, ils ont été fournis par des ingénieurs passionnés qui ont supervisé les activités liées à la construction de barrages. Cet article met en lumière cinq ingénieurs sélectionnés qui ont contribué de manière significative à l'ingénierie des barrages suisses. Leur rôle au sein du Comité national suisse des grands barrages est bien sûr pertinent.

## 1 INTRODUCTION

Although hydropower remains the largest renewable electricity technology by capacity and generation, the current capacity growth trends are insufficient to place it on the trajectory under the Net Zero Scenario. Reaching about 5 700 TWh of annual electricity generation by 2030 will require a 3% average annual generation growth between 2021 and 2030, which appears additionally challenging when considering the accelerating disturbances to the water availability caused by the climate change and an ageing fleet of hydropower plants. On the capacity side, an average of 50 GW of new hydropower plants needs to be connected to the grid annually until 2030, which is more than twice the average of the past five years. Much greater efforts, especially in developing and emerging markets, will be globally required to achieve that pace of growth.

Thanks to its topography, geology and high levels of annual rainfall, Switzerland has ideal conditions for the utilization of hydropower. Starting in 1892 with the first plant for electricity generation installed in St. Moritz, hydropower underwent an initial period of expansion, and between 1945 and 1970 it experienced a genuine boom during which numerous new power plants and low-head dams were commissioned in the Swiss Plateau, together with large-scale storage plants and reservoir dams in the Alps.

Based on the estimated mean production level, hydropower still accounted for almost 90% of domestic electricity production in 1970, but this figure fell to around 60% by 1985 following the commissioning of Switzerland's nuclear power plants, and is now around 57%. Hydropower therefore remains Switzerland's most important domestic source of renewable energy.

Today there are nearly 700 hydropower plants in Switzerland having a capacity of at least 300 kW, producing an average of around 37,260 GWh/y, 48% of which is supplied both by run-of-river power plants, and by storage power plants, whereas 4% by pumped storage power plants (from natural inflow only). 63% of hydroelectricity are generated in the mountain cantons of Uri, Grisons, Ticino, and Valais, while Aargau and Bern also generate significant quantities. 11% of Switzerland's hydropower generation comes from facilities located on water bodies along the country's borders. The hydropower market is worth around 1,8 billion Swiss francs and is therefore an important segment of Switzerland's energy industry (Bundesamt für Energie 2021). Figure 1 shows a top view of the beautiful Hongrin Dam and its Veytaux emergence structure located on the shores of Lake Geneva.

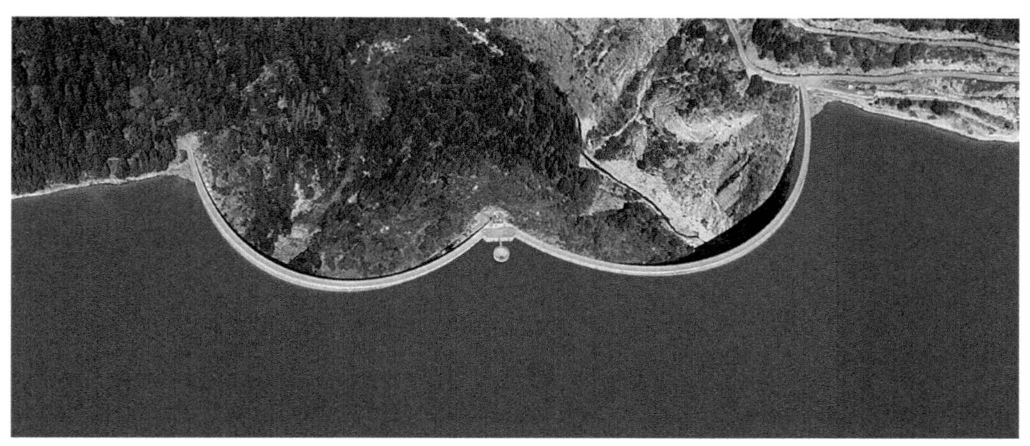

Figure 1. (up) Hongrin Dam in Canton of Vaud, (down) Veytaux I emergency outlet into Lake Geneva https://www.alpiq.ch/energieerzeugung/wasserkraftwerke/pumpspeicherkraftwerke/hongrin-leman; https://www.raonline.ch/pages/edu/nw3/power01a4a7.html.

This paper deals with engineers at the heart of the early dam engineering knowhow in Switzerland. Five selected portraits are provided who mostly have a close relation to the Swiss National Committee on Large Dams. In addition to the main projects realized, their input to the Swiss engineering organizations is provided, and their roles as technical and human leaders is highlighted.

## 2 THE SWISS COMMITTEE ON DAMS

The Swiss National Committee on Large Dams (SNCOLD) is a private association representing the Swiss dam community within the International Commission on Large Dams (ICOLD). The committee's objective is to promote construction, operation, maintenance and monitoring of hydraulic structures and their environment. To achieve this goal, it unites specialists from various branches of dam technology offering them a platform to discuss experiences, publish technical papers and organize symposia and workshops related to dam engineering.

Based on a loose union of five Swiss dam engineers, the Swiss Dam Commission was established in 1928. On December 20, 1948, Henri Gicot chaired the founding assembly of the SNCOLD. It comprised 68 members, many of them from industry and electric utilities. During the 1950s, important dams were built, among which Grande Dixence (PG), Mauvoisin (VA), Moiry (VA), Göscheneralp (ER), Valle di Lei (VA), Albigna (PG), Luzzone (VA) or Malvaglia (VA). Typical topics of discussion were then the vulnerability under military attacks or frost resistance of mass concrete.

Switzerland counts some 1200 dams, of which most are small, however. About 225 dams are under the direct supervision of the Swiss Federal Office of Energy in view of their dimensions and the potential dangers they may include. 162 dams meet the criteria for large dams established by the International Commission on Large Dams (ICOLD).

Based on the number of dams under federal supervision, the dam density in Switzerland is more than 5 dams per 1000 km$^2$. Most of these dams are key elements of major hydropower schemes. 80% of the dams are in mountainous regions. The construction of the schemes was vital for the development of these economically less favorable areas.

Dam construction in Switzerland has a long tradition. The oldest dam under federal supervision is the 15 m high Wenigerweiher embankment in the city of St Gallen. It was constructed in 1822 for the energy supply of nearby industry and currently is still impounded and in use, serving ecological purposes nowadays by creating an amphibian spawning area of national importance. Most of the dams taken into service in the 19$^{th}$ century were located close to cities because it was either prior to the invention of electricity or then difficult to transfer electricity over long distances. An example is the Maigrauge Dam, a roughly 20 m high gravity dam built in 1872 on the Sarine River slightly upstream of Fribourg city. The dam is currently also still in operation for hydropower use and has greatly contributed to the economic development and the prosperity of the city.

The first dams in the Alps were built starting from 1900. This includes the 112 m high Schräh Dam commissioned in 1924, then being the highest dam worldwide. The real start of Swiss dam development initiated after WWII, however. 86 large dams were commissioned between 1947 and 1970, four of which being higher than 200 m:

- Mauvoisin Arch Dam (1957), 237 m high, raised to 250 m in 1990;
- Grand Dixence Dam (1961), 285 m high, still the highest gravity dam worldwide;
- Luzzone Arch Dam (1963), 208 m high, raised to 225 m in 1998;
- Contra Arch Dam (1965), 220 m high.

The large Swiss dams are mainly concrete structures; only the Göscheneralp Dam (155 m) and the Mattmark Dam (117 m) are rockfill dams higher than 100 m, followed by the 91 m high Marmorera earthfill dam.

Some 90% of the technically feasible hydropower potential is currently in operation. The remaining 10% is difficult to achieve for economic reasons and because of increased conflicts of interests. One of the major challenges of current dam engineering in Switzerland is to cope with aging of the large fleet of existing dams at the highest level of safety. The Swiss Committee on Dams (2011) portrays the main Swiss dams including the name, the owner, the dam purpose, the type of foundation, various technical data, a short history, and further technical specifications of interest. Each dam is also described with a plan view and a cross-section, plus its location in Switzerland, and a full-page photo.

The main purpose of this paper is the presentation of five outstanding past Swiss dam engineers, who have greatly contributed to the technical development of Swiss dams and its electricity potential. These personalities include:

- Guillaume Ritter, hydraulic engineer and designer of the first concrete dam in Europe at Maigrauge on the Sarine River
- Heinrich Eduard Gruner as a pioneer in dam engineering, realizing together with Alfred Stucky the Montsalvens Dam as the first arch dam of Europe
- Alfred Stucky, professor of hydraulic engineering at the Ecole Polytechnique Universitaire de Lausanne (EPUL) and founder of its hydraulic laboratory, with publications for the design of dams
- Henri Gicot, designer of many arch dams and the first dams with either parabolic or elliptic arches

– Giovanni Lombardi introducing the Lombardi slenderness coefficient characterizing an arch dam, and criteria for dam cracking.

## 3 PORTRAITS OF SWISS DAM ENGINEERS

### 3.1 *Guillaume Ritter*

Ritter was born in Neuchâtel in 1835 to Alsatian parents; he died in Monruz near Neuchâtel in 1912 (Anonymous 1913; Walter, 1977; Vischer 2001). After basic schooling in Neuchâtel, he entered *Ecole Centrale des Arts et Manufactures* in Paris, graduating in 1856 with a diploma as *Ingénieur constructeur* (civil engineer). Subsequently, he devoted himself in Neuchâtel specifically to projects of urban water supply and industrial hydropower. One of the extraordinary projects dealt in 1872 with the hydroelectric power plant on Sarine River near Maigrauge (German: Magerau).

Ritter was considered an enthusiastic and inspiring innovator. The plant near Maigrauge was the second major Swiss power plant after that on the Rhine in Schaffhausen. It supplied Fribourg with a network of rope transmissions and a pressurized water network. Electrification did not take place until 1891-1895. With a height of 21 m, a crown length of 195 m and a dam capacity of 1 million m$^3$, the Sarine Dam was a novelty in Switzerland. Ritter was the first in Europe to use concrete as a building material, after it had been used for the first time for American dams just a few years earlier. His financially and technically most complex proposal did not get beyond the planning and acquisition stage. It was the water supply of Paris from Lake Neuchâtel, involving a 37 km base tunnel under the Swiss Jura and a 470 km long pipeline. Ritter expected construction to begin in 1900, the need of 400 million Swiss francs in investment costs by financing until the year 2000.

### 3.2 *Heinrich Eduard Gruner*

He was born in 1873 in Basel, passing away there in 1947. His family was a true engineering household. He obtained the civil engineering diploma from ETH Zurich, undertook then study tours to Saxony, England, and the USA, starting his engineering career in 1914 at his father's office. His first project involved the Laufenburg Power Plant on the Rhine River upstream of Basel, where he headed the local works. He there collected his first practical experiences, applying these later to similar projects including the Eglisau and the Ryburg-Schwörstadt low-head dams, both also located on the Rhine River (Hager 2003).

He and his collaborator Alfred Stucky designed around 1920 the first Swiss arch dam at Montsalvens in Canton Fribourg, of 55 m height. Gruner also realized the importance of the scientific hydraulic modelling, being one of the financial supporters of the ETH Hydraulic Laboratory taken into service in 1930 under its director Prof. Eugen Meyer-Peter (Hager et al. 2021). Later, Gruner's main activities were in Iran, improving there the irrigation techniques, and in Egypt as a member of the Commission for the second Aswan Dam. He was in addition the first Swiss delegate of ICOLD. Gruner was awarded the honorary doctorate from ETH Zurich in 1930, and was since 1925 a member of the American Society of Civil Engineers (ASCE). Mommsen (1962) gives a detailed and historically interesting account of the Gruner engineering family.

3.3 *Alfred Stucky*

He was born in 1892 in La Chaux-de-Fonds (NE) and passed away in Lausanne (VD) in 1969. Stucky was a true expert in hydraulic engineering during the golden ago of hydropower. After having graduated at ETH Zurich in 1915, he joined the engineering office of Gruner in Basel. In parallel, he submitted to ETH Zurich in 1920 a PhD thesis on arch dams. He was appointed hydraulic engineering professor at EPUL in 1926. He founded the EPUL Hydraulic Laboratory in 1928 and initiated his private engineering office in parallel to his commitment at EPUL. Stucky further acted as EPUL president from 1940 to 1963, when retiring from EPUL and concentrating on consulting work until his passing (Hager 2003).

Stucky's career was threefold as hydraulic engineer, researcher in dam engineering and as the organizer of the engineering school. Based on his PhD, he initiated the design of arch dams in Switzerland. He was an expert of the Italian Gleno Dam disaster of 1923 (Stucky 1924). Stucky also investigated sea wave forces on vertical walls in his laboratory, based on his international expertise. In 1936, his interest into surge tanks started, culminating in a text book (Stucky 1962).

He was in addition largely involved in the Grande Dixence Dam, currently still the largest gravity dam worldwide. He also was for 20 years president of the technical journal *Bulletin Technique de la Suisse Romande*. He was awarded the honorary doctorate degree from ETH Zurich in 1955, among many other distinctions for his outstanding engineering career.

3.4 *Henri Gicot*

He was born in 1897 in Le Landeron (NE) passing away in 1982 in Fribourg. After having completed his civil engineering studies at ETH Zurich in 1919, he joined as previously did

Stucky the engineering office Gruner in Basel. From 1927 to 1971, he owned his engineering office in Fribourg. In addition, Gicot was the president of the SNCOLD from 1948 to 1961, and from 1953 to 1967 a member of the ETH Board. He was in addition an expert of the World Bank active in Asia and South America.

Gicot became internationally known as a dam engineer for the Rossens Dam built on Sarine River from 1944 to 1948. He was in addition also responsible for the Montsalvens Dam close to Broc (FR) built on the Sarine River from 1919 to 1921, the Gebidem Dam below the Aletsch Glacier from 1949 to 1950, the Delcommune Dam in the Belgian Congo from 1950 to 1952, the Vieux Emosson Dam from 1954 to 1955, Zeuzier Dam from 1955 to 1957, Schiffenen Dam from 1960 to 1964, and the beautiful Hongrin Dam built from 1964 to 1968 (Figure 1). He was awarded the honorary doctorate degrees from the University of Fribourg in 1962, and of ETH Zurich in 1968 (Schnitter 1982).

### 3.5 *Giovanni Lombardi*

He was born in 1926 in Lugano (TI), passing away in 2017 in Monte Carlo, Monaco. Lombardi was an internationally renowned civil engineer who was an outstanding expert in tunnel and dam projects. He graduated in 1948 from ETH Zurich as a civil engineer obtaining in 1952 the PhD title with a work on slender arch dams. He founded in 1955 with G. Gellera the engineering office Giovanni Lombardi PhD Consulting Engineers in Lugano, from 1989 a stock corporation and renamed in Lombardi AG Consulting Engineers in Minusio (TI).

His firm's projects included numerous tunnels, such as the Gotthard Road tunnel or the Gotthard Base Tunnel. Other projects were dams in the Verzasca Valley in 1965, also known as Contra Dam, and in the Valle Morobbia (Lago di Carmena), dams in Austria (Kops, Kölnbrein), Italy (Ridracoli Dam in Emilia-Romagna, Flumendosa Dam in Sardinia) or in Mexico (210 m high Zimapan Dam). He was a member of the commission to investigate the subsidence that occurred in 1978 at Zeusier Dam, caused by the neighboring construction of an exploratory tunnel for a road tunnel in which water ingress occurred, draining the groundwater in the surroundings. From 1979 to 1985 he was president of the Swiss Committee on Dams. He was the first Swiss president of ICOLD from 1985 to 1988. In 2008 he received the Swiss Award, a prize for outstanding Swiss personalities. He holds honorary doctorates from EPFL (1986) and the *Politecnico di Milano* (2004).

## 4 CONCLUSIONS AND OUTLOOK

Switzerland's history of modern reservoir dams is 200 years old. Starting from small embankments, mainly low-head hydropower dams along rivers were erected in the 19th century. These were located close to industries because the transmission of electricity remained a problem until the end of the century. Switzerland, a country without notable treasures of the soil, had to concentrate early on energy obtained from hydropower. As an example, most trains were driven from WWI with electricity, or the so-called white coal, given the absence of conventional black coal. The large dams were erected after WWII, with dams reaching nearly a height of 300 m. This intense phase of dam construction activity came to its end around 1970, given that most of the favorable dam sites had been used.

Switzerland had and still has many outstanding dam engineers, who explored the possibilities to construct appropriate dam structures to supply electricity and protect against floods. While the focus was on Switzerland until the 1980s, it has since then shifted to dam engineering worldwide (Droz 2023), enlarging the scope of the dam purposes also to irrigation and water supply. Still today, some 50% of the electricity consumed in Switzerland originates from hydropower. The present paper presents five selected individuals having greatly contributed to Swiss dam engineering. They represent all major parts of the country and have been active both in their homeland and internationally. A short review of their education is provided, along with the main projects of their professional activity. In addition, important other occupations are presented, such as in national or international committees, particularly within the Swiss National Committee on Large Dams, who has reached its 75th anniversary. It is also noted that the Swiss dam engineering will have an active role in the future, given its relevance for water resources management and society, not least in view of climate change adaptation and new demand for multipurpose reservoirs (Boes & Balestra 2023).

## ACKNOWLEDGEMENT

The author would like to acknowledge the interest of his colleague Prof. Dr. Robert M. Boes, Director VAW, in the present work, by adding suitable comments.

## REFERENCES

Anonymous. 1913. Guillaume Ritter. *Verhandlungen der schweizerischen naturforschenden Gesellschaft*, 96: 28–33 (in French).

Boes, R.M., Balestra, A. 2023. Dams in Switzerland. Proc. ICOLD European Club Symposium *Role of dams and reservoirs in a successful energy transition* (Boes, R.M., Droz, P. & Leroy, R., eds.). Taylor & Francis, London.

Bundesamt für Energie 2021. *Wasserkraft Schweiz*: Statistik 2021 https://www.bfe.admin.ch/bfe/de/home/news-und-medien/medienmitteilungen/mm-test.msg-id-88652.html.

Droz, P. 2023. Swiss dam engineering in the world. Proc. ICOLD European Club Symposium "Role of dams and reservoirs in a successful energy transition" (Boes, R.M., Droz, P. & Leroy, R., eds.), Taylor & Francis, London.

Hager, W.H. 2003. *Hydraulicians in Europe* 1800-2000. IAHR: Delft, the Netherlands.

Hager, W.H., Schleiss, A.J., Boes, R.M. & Pfister, M. 2021. *Hydraulic engineering of dams*. CRC-Press, Taylor & Francis Group: Boca Raton, London, New York.

Mommsen, K. 1962. *Drei Generationen Bauingenieure* (The generations of civil engineers). Schwabe: Basel (in German).

Schnitter, G. 1982. Henri Gicot. *Wasser, Energie, Luft*, 74(9): 255 (in German).

Stucky, A. 1924. La rupture du barrage du Gleno (Dam break of Gleno Dam). *Bulletin Technique de la Suisse Romande*, 50(6): 65–71; 50(7): 79-82; 50(9): 107-108 (in French).

Stucky, A. 1962. *Druckwasserschlösser von Wasserkraftanlagen* (Surge tanks of hydropower plants). Springer: Berlin (in German).

Swiss Committee on Dams 2011. *Dams in Switzerland*: Source for worldwide Swiss dam engineering. Buag: Baden-Dättwil (in German).

Walter, F. 1977. Ritter Guillaume. *Encyclopédie du Canton de Fribourg*, 2, 497. Office du livre: Fribourg (in French).

Vischer, D.L. 2001. *Wasserbauer und Hydrauliker der Schweiz* (Swiss hydraulic engineers and hydraulicians). Verbandsschrift 63. Schweizerischer Wasserwirtschaftsverband, Baden (in German).

*Role of Dams and Reservoirs in a Successful Energy Transition – Boes, Droz & Leroy (Eds)*
*© 2023 The Author(s), ISBN 978-1-032-57668-8*

# Water resources optimisation – A Swiss experience

R. Leroy
*Alpiq SA; Swisscod - Swiss Committee on dams*

ABSTRACT: Historically, during the 20th century and up to the present day, water management has evolved significantly. During this period, much progress has been made for better understanding the importance of water as a vital resource and in implementing appropriate measures for its management and arbitration. During the 20th century, hydraulic infrastructures such as dams and reservoirs were built to store and distribute water more efficiently. Climate change and overexploitation of resources make water management an increasingly complex issue.

RÉSUMÉ: Depuis toujours, durant le 20e siècle et jusqu'à nos jours, la gestion de l'eau a évolué de manière significative. Au cours de ces périodes, de nombreux progrès ont été réalisés dans la compréhension de l'importance de l'eau en tant que ressource précieuse et dans la mise en place de mesures pour sa gestion et son arbitrage. Au cours du 20e siècle, des infrastructures hydrauliques, telles que les barrages, les réservoirs ont été construites pour stocker et distribuer l'eau de manière plus efficace. Les changements climatiques et la surexploitation des ressources font de la gestion de l'eau un enjeu toujours plus complexe.

## 1 INTRODUCTION

Switzerland, a country of mountains, streams, rivers and lakes, possesses the most beautiful, the most priceless of all riches: blue gold! A precious reserve of life and energy in the heart of Europe. Due to its altitude and the Alpine foothills, more than two thirds of the precipitations are retained in winter in solid form. In spring and summer, snow and ice turn into liquid. From time immemorial, man has tried to tame these streams of life. He has fought against water and its overflow. Above all, he has fought for water and its benefits. The history of the Alpine cantons bears witness to these struggles. It is the patient work of damming up the rivers, the extraordinary network of irrigation systems and bisses and the mastery of this energy.

Water management is an important issue. While ensuring long-term sustainability, an integrated approach allows for equitable management and sharing of water resources between competing uses such as hydropower generation, irrigation, flood protection, drinking water supply and protection of aquatic ecosystems.

The last century was the epic of dam construction. The titanic struggle in the heart of the mountains to capture and store the most incredible energy: hydroelectricity. A clean, renewable, ecological energy. The production of hydroelectric energy in Switzerland is very important, as it represents nearly 60% of the country's electricity production. Storage dams are one of the key facilities, as they allow water to be stored for later use, whether for hydroelectric production, to support low-water flows, or to regulate river flows. The large Swiss dams, such as the Maggia scheme, the Grimsel scheme, the Mattmark embankment dam, the Grande Dixence dam or the Emosson dam, are examples of dams that contribute and will continue to contribute significantly to electricity production in Switzerland.

At all times, when it has been a question in the past of catching and transferring water, today of making a better use of water in a difficult energy context, and tomorrow, considering the predicted climate changes and the necessary arbitration between the various stakeholders, resource management and optimisation have been, are and will be the challenges, both on the technical and societal aspects.

## 2 CASE STUDIES

Water. . .the original element without which no form of life is possible.
Water. . . Silent and calm, slumbering in the reservoirs.
Water. . . Wild and tumultuous, raging through the turbines.
Water. . . an inexhaustible source of energy

### 2.1 *Grande Dixence*

One of the most striking episodes in the conquest of this "white coal" is without doubt the construction of the Grande Dixence complex. This pharaonic construction site is a jewel of ingenuity and human courage to develop a unique glacial basin of 350 $km^2$.

Figure 1. Grande Dixence gravity dam.

Since the end of the Second World War, Switzerland has needed energy to meet the demands of its industrial development. In 1945, the Swiss Federal Water Board drew up a comprehensive inventory of the country's hydroelectric potential. Analysing the possibilities in the Rhone basin, the experts came to the conclusion that there were still a number of valleys that could be exploited under economically attractive conditions. The Val des Dix soon emerged as the site with the greatest development potential. This high valley had the ideal geological and topographical conditions to become a giant reservoir. No human settlements were affected; the only agricultural land was highland pasture, and above all, the predicted storage capacity was enormous: 400 million $m^3$.

Geologists, hydrologists, topographers and engineers set out to solve two major problems. On the one hand, to enlarge the existing lake with the Dixence complex built some fifteen years earlier. On the other hand, to create an catchment network capable of collecting water from the neighbouring valleys of Mattertal, Ferpècle and Arolla. More than three thousand men fought this battle until the beginning of the 1960s. A daring and avant-garde project which today contributes to the well-being of the community.

However, although the sight of this gravity dam, the highest in the world, is impressive, it is only the tip of the iceberg. The particularity of Grande Dixence scheme, and the genius of its

Figure 2. Grande Dixence dam under construction.

Figure 3. Grande Dixence dam under construction.

designers, is that it is able to collect the water from 35 glaciers, from the confines of the Zermatt valley to the Hérens valley. To bring all this water to the Val des Dix, men had to drill the rock before excavating it to create the 75 water intakes and water transfer gallery network of around 100 kilometres with a gradient of 2 ‰ along its length. Some important glaciers such as Ferpècle, Arolla, Z'Mutt and Gorner are located lower than the level of the main collector at an altitude of 2 400 metres. Thus, 4 pumping stations are needed to transfer water in the main reservoir.

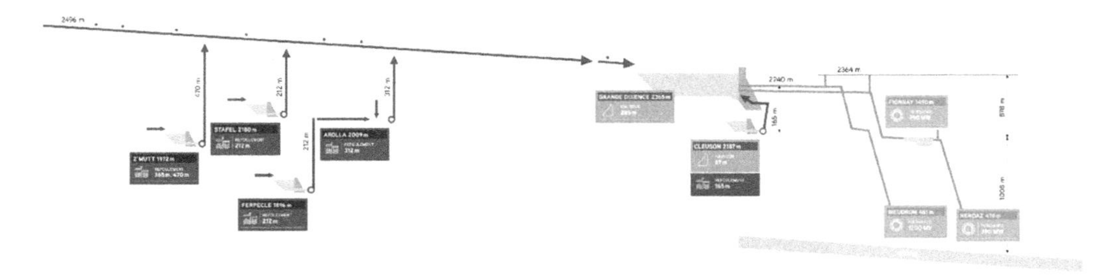

Figure 4. Grande Dixence scheme - longitudinal profile.

Collecting, pumping, and bringing the water from 35 glaciers to the Val des Dix, in order to use it for power generation, has been a difficult and hazardous task. In recognition of the commitment of previous generations, the real challenge is nowadays to carefully manage this source of energy and to fully exploit its value. It is a question of finding the best possible match between the constraints of the installation, the power demands, and the market prices. At first sight, the problem is simple: fill the Lac des Dix with as much water as possible during the limited period of glacial and snowmelt. But in reality, it is a headache: it is necessary to take into account the capacity of the water transfer galleries, but above all of the main collector, to integrate the variations in the flows of the water intakes, to monitor the meteorological forecasts and the forecasts of short, medium and long term power demand, to carry out the pumping of the water collected in the neighbouring valleys during the favourable hours, without forgetting to return large quantities of water for ecological, tourist and contractual reasons.

Figure 5. Grande Dixence scheme- 3D view.

The hydro-scheme is there to provide high quality energy to supply the market at peak times. The level of storage in the scheme must be optimised for maximum availability before periods of high demand. To achieve a satisfactory balance, the optimisation of the summer pumped storage energy must be done in relation to the energy produced in the winter by integrating a multitude of parameters. All the data collected allows for optimised management of the inflows and outflows required to fill the reservoir. The challenge of this high-precision management is crucial: one million cubic metres of water lost represents more than four million kWh of energy in winter. The adaptability of the scheme and its development to the changing contexts is to be noted. The Cleuson-Dixence development with the Bieudron 1200 MW power station corresponds to a quest for power and stabilisation of the power grid.

Figure 6. Bieudron 1200 MW powerplant.

Climate change increases the multiple threats to water availability. The spatial and temporal pattern of precipitation and water availability is changing. The frequency, intensity and severity of extreme weather events are likely to increase. Glaciers are melting and natural water storage is diminishing. In view of this, the project to build a new reservoir, the Gornerli reservoir in the upper Zermatt valley, is intended to secure the valley by reducing flooding and providing a new strategic winter reserve.

## 2.2 *Emosson*

At the beginning of the 20th century, engineers had already noticed that the Emosson site was suitable for the construction of a large hydraulic reservoir, but because of the limited natural water inflows and the impossibility, at the time, of collecting water via long galleries and of carrying out large-scale pumping, the project was not developed until 1953. The preliminary project to use the waters of the Drance d'Entremont, the Drance de Ferret, the Trient and the Eau Noire to fill this new retention structure in the Barberine region was launched and discussions were held to obtain the necessary concessions to carry out this work. At that time, Electro-Watt, which was responsible for the construction of the Mauvoisin scheme and was awarded the Mattmark water concession in February 1954, abandoned the "Grand Emosson" project.

Motor Columbus SA d'Entreprises Electriques de Baden founded the company "Usines hydroélectriques d'Emosson SA" in 1954, which became "Electricité d'Emosson SA" in 1967. The

Figure 7. Gorner glacier valley - view from above.

Figure 8. Photomontage of the future Gornerli dam.

acquisition of the local community concessions and the extension of the project into French territory took place at the same time and led to the admission of Electricité de France into the company in 1955. It took many years to get the administrative apparatus of both countries up and running and to get the project underway. The project's interest in the highly sought-after peak energy quality was undisputed. At the same time as the project was being studied, discussions were initiated with the Swiss Federal Railways (SBB), whose previously used Barberine reservoir would be flooded. France and Switzerland decided to change the border

Figure 9. Emosson dam under construction.

line. The wall of the Emosson dam would have been cut in the middle by the Franco-Swiss border, leaving its right side on French soil and the rest of the structure in Switzerland, while the Châtelard-Vallorcine hydroelectric power station would have been entirely in Swiss territory. The agreement of August 23, 1963 made it possible to rectify the Franco-Swiss border so that the project could be carried out in accordance with the respective interests of the two States. France and Switzerland proceeded to an "exchange of territory of equal area" which placed the wall of the Emosson dam entirely in Switzerland and, symmetrically, the Châtelard-Vallorcine hydroelectric power station entirely in France. Works began in 1967 and the plant was commissioned in 1975. The constructions, more than 65% of which are in Switzerland and 35% in France, constitute an indivisible entity. The authorities of the two countries, meeting in the permanent supervisory commission, are also very diligent in dealing with the problems that arise due to the existence of national borders in the middle of the installations and the different legislation between the two countries.

Figure 10. Emosson dam.

The Franco-Swiss Emosson scheme drains water from the French valleys of the Arve and Eau Noire and Swiss water from the Val Ferret and the Trient Valley. The South Collector, 8.55 km long, collects water from the Lognan, Argentière and Tour glaciers, which flows towards the dam gravitationally, via a siphon. The West Collector, 7.95 km long, collects the water from the Bérard and Tré-les-Eaux valleys at an altitude of 1 990 m above sea level and conveys it directly into the dam's reservoir. The East Collector collects water from La Fouly at an altitude of 1 550 m above sea level. The water comes from the Val Ferret, the Saleinaz and Trient glaciers and various other torrents. The gallery is 18.3 km long. This water flows into the Esserts basin. It is either turbined in Vallorcine or pumped into the Emosson reservoir.

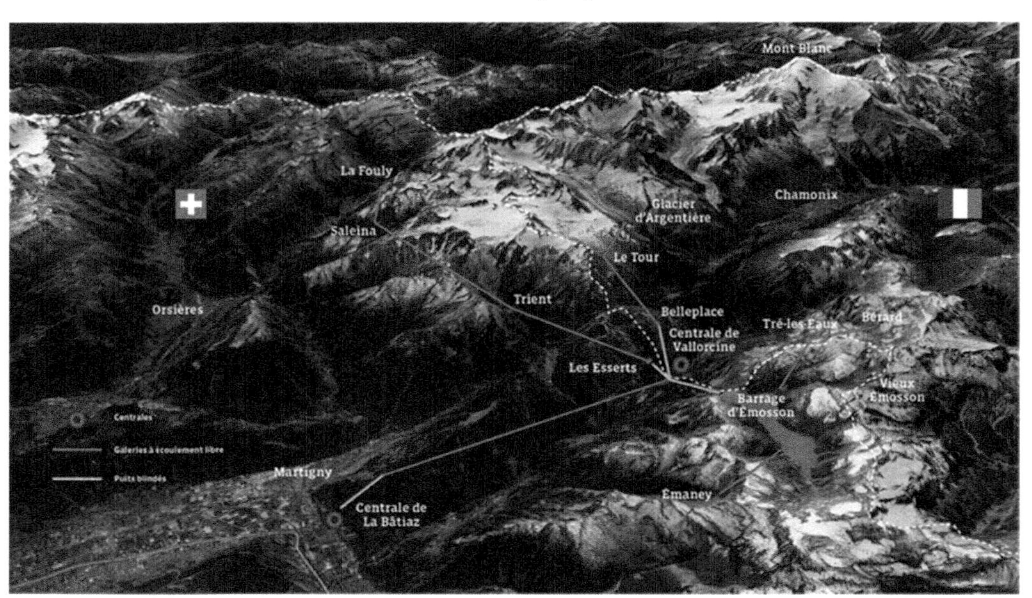


Figure 11. Emosson scheme - 3D view.

Figure 12. Nant de Drance PSP – heightening of the Vieux-Emosson dam – upper reservoir.

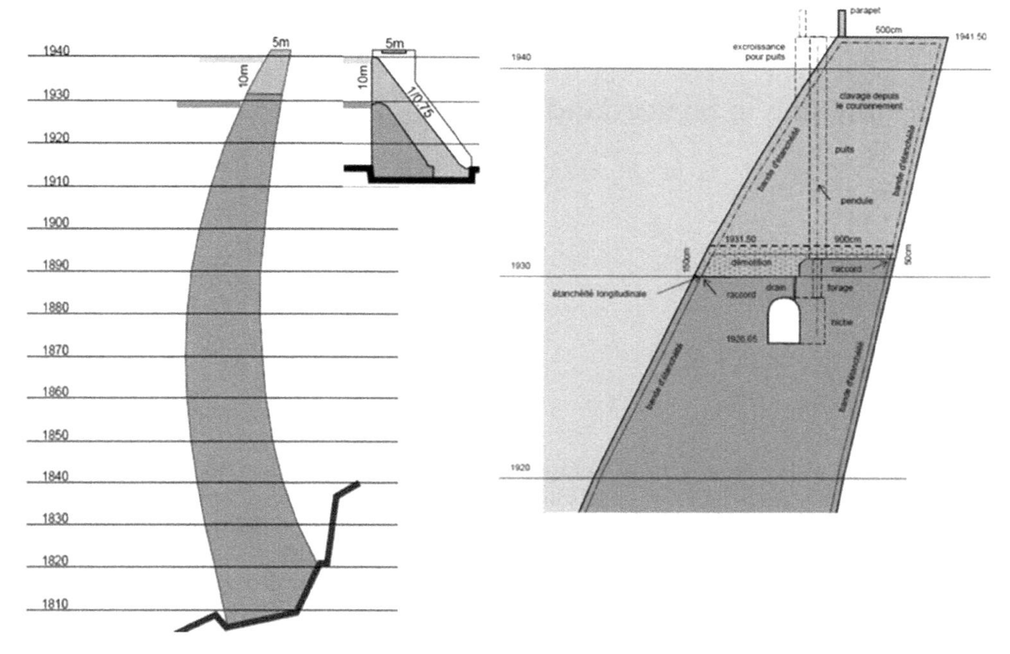


Figure 13. Emosson dam heightening project – lower reservoir.

The adaptability of the scheme to changes is to be noted. The Emosson scheme responds also to a constant quest for power and stabilisation of the power grid. The evolution of the very volatile energy context has allowed the construction of the Nant de Drance pumped storage power station to cope with the intermittence of renewable energies. This project, commissioned in 2022, required the raising of the Vieux-Emosson dam, built in 1956 for the Swiss Federal Railways (SBB), its reservoir serving as the upper basin, the lower basin being that of Emosson.

A project to raise the Emosson dam is also under study.

## 3 CONCLUSIONS

As the two case studies presented demonstrate, the optimisation of water resources is very multifaceted. In the past, as well as today and in the future, societal issues have been, are and will be the main development objectives of these large-scale schemes. These experiences can naturally be called upon to evolve and be applied in other geographical contexts.

Role of Dams and Reservoirs in a Successful Energy Transition – Boes, Droz & Leroy (Eds)
© 2023 The Author(s), ISBN 978-1-032-57668-8

# Dam heightening in Switzerland

A. Wohnlich
*Gruner Stucky SA*

A. Fankhauser
*Kraftwerke Oberhasli AG*

B. Feuz
*Independent Engineer*

ABSTRACT: Raising the height of existing dams is becoming an increasingly topical issue in Switzerland. This particular construction method offers a number of environmental, technical and economic advantages, which are presented in this article. Swiss engineers have been raising their dams for more than a century, and the article describes the various structures involved. Finally, a perspective on future dam heightening is offered considering the climate and energy issues prevailing at the start of the 21st century.

## 1 INTRODUCTION

The technology of dam heightening is not recent. In fact, the first known heightening of a dam in Switzerland during the modern era of dam construction (from the end of the 19th century) dates back to 1910 (Maigrauge dam, canton of Fribourg). Throughout the 20th century and into the 21st century, no less than a dozen dams have been raised in Switzerland. Most of these were concrete dams, and the vast majority were gravity dams.

This article begins by outlining the advantages of raising dams and the main issues involved in this challenging work. The main aim is to increase the storage volume of the reservoir. However, this is not always the only objective, as will be shown below.

The various heightened dams are presented, highlighting the characteristics and specific features of each structure. The first part of the publication presents mainly and chronologically heightened gravity dams, all of which are of moderate size. The second part of the publication deals with the more recent and much larger scale raising of arch dams. Over the past 30 years, three major arch dams have been raised in Switzerland: Mauvoisin, Luzzone and Vieux Emosson dams.

Swiss know-how in the field of dam heightening has also been exported: Swiss engineering has had the opportunity to make its mark from time to time, as will be recalled in a dedicated section.

Finally, considering the tense geopolitical situation prevailing in Switzerland and Europe at the start of the 21st century, which has direct repercussions on the energy sector, as well as the effects of global climate change, a perspective is offered in conclusion, showing that the interest in raising existing dams is likely to continue over the coming decades.

## 2 INTEREST OF DAM HEIGHTENING

The main purpose of raising a dam is to create additional storage volume in the reservoir. The possibility of having a greater storage volume offers major economic advantages to the operator of the structure, in the form of increased flexibility in the management of its hydroelectric scheme (optimization of productive hours, as well as transfer of part of the summer production to the winter).

Considering the shape of a valley used as a reservoir, the storage volume is generally always significantly higher at the top of the dam than in the talweg. Each additional metre of height therefore contributes more to the volume stored than the levels below.

In contrast, when considering the shape of concrete dams, it can be seen that the greatest dam thickness is close to the foundations, and that as the dam rises, it becomes thinner. The volume of concrete required to raise a dam is therefore relatively small compared with the additional volume of water stored in the upper levels. In principle, this twofold geometric advantage makes dam heightening projects economically attractive.

In addition, the environmental issues involved in raising a dam are normally reduced compared with building a new dam on a greenfield site. When a dam is raised, the site has already been constructed for many years and public acceptance is easier, as the impact on the environment and on the landscape is reduced.

What's more, in most cases (most heightening works are carried out on the top of the structure itself, or on its downstream side, only rarely on its upstream side), a heightening project can be carried out without emptying the existing reservoir and therefore without any loss of operation, which is a key economic factor for the operator of the hydroelectric scheme.

Finally, it should be noted that raising a dam is generally only possible if it can take advantage of 'generous' dimensioning at the time the dam was designed, developments in science (better knowledge of construction materials, characteristics of rock foundations, etc.), and progress in design tools (more recent and sophisticated calculation methods).

## 3 HISTORICAL REVIEW

The review presented below spans the whole of the 20$^{th}$ century up to the 2020s. The main group of heightened dams described first below concerns only relatively modest-sized gravity dams, between 10 m and 40 m high. This is clearly the type of structure that has been most widely considered for raising during the 20$^{th}$ century.

A second group of gravity dams is dealt with separately. These are two dams (Muslen and List) for which raising the water level was not the primary aim but was subordinate to the objective of strengthening the structure. As part of such a safety project, it turned out that it was also possible to take advantage of major reinforcement work to raise the dam.

Finally, a third group of dams is discussed next. These are two relatively high structures (Salanfe and Les Toules), between 50 m and 90 m, which were raised very shortly after the construction of the initial dam, and for which the raising was designed from the outset. The reason for the two-stage construction was the operator's intention to bring its hydroelectric scheme into service as soon as possible. For various reasons, the hydraulic circuit and the hydroelectric powerplant were operational earlier than the dam, and the most economical way of commissioning the scheme as quickly as possible was to build a small retaining structure sufficient to supply the water intake, and then to plan the subsequent construction of the dam to its full height in a second phase.

The presentation and description of the three large arch dams raised between 1990 and 2020 is addressed separately in Chapter 4.

### 3.1 *Gravity dams*

The structures described below are presented by date of heightening.

Maigrauge dam (Figure 1, left) is one of the oldest dams in Switzerland. It was built in 1870-72 and heightened in 1909-10, probably because of the rapid silting up of the reservoir. It should be noted that the prestressed anchors (1600 kN/2 m) shown in the cross-section in Figure 1 were added in 2000-03 to ensure that the structure complies with stability requirements (structural safety). Relatively high tensile stresses could develop at the heel of the dam, particularly in the event of exceptional loads such as the safety flood or an earthquake.

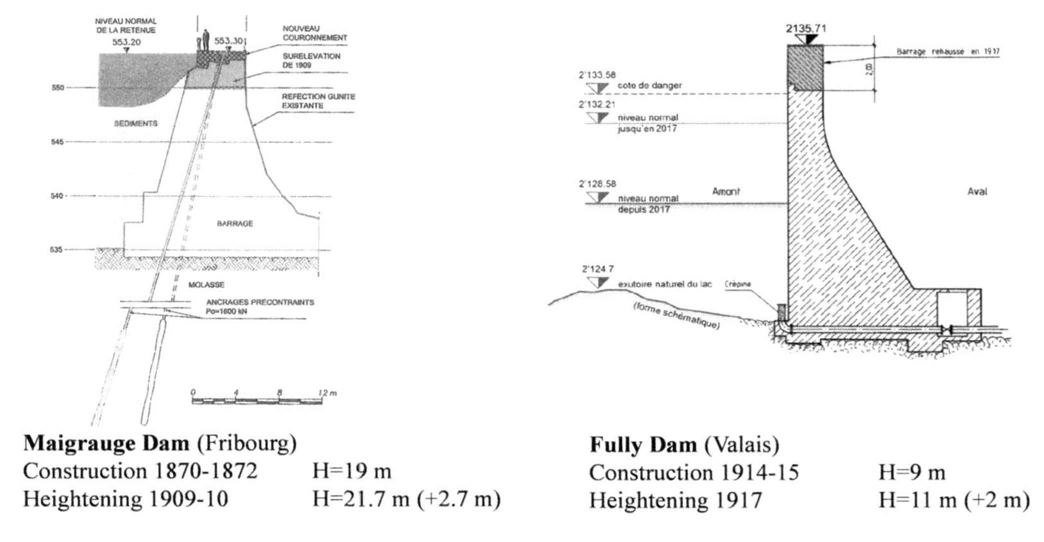


Figure 1. Maigrauge dam and fully dam.

Fully dam (Figure 1, right) was built in 1914-15 and raised by 2 m in 1917. The dam axis is not straight, but curved in plan, which probably generates a three-dimensional effect and explains how the raised section can be stable. It should be noted, however, that the dam has been operated at a lower water level for several decades.

Lago Bianco South dam (Figure 2, left), built at the beginning of the 20th century, was raised by 4 m in 1941-42. The dam axis also has a curvature in plan, which in reality makes it an arch-gravity dam. Over time, the concrete of the raised section was found to suffer from the alkali-aggregate reaction (AAR), resulting in vertical cracking and increased deformation. The raised section was rehabilitated in 2000-01.

It should also be noted that as part of the development of the pumped-storage project, a new raising of 3.5 m was studied in 2011-12. However, this project has not yet been implemented.

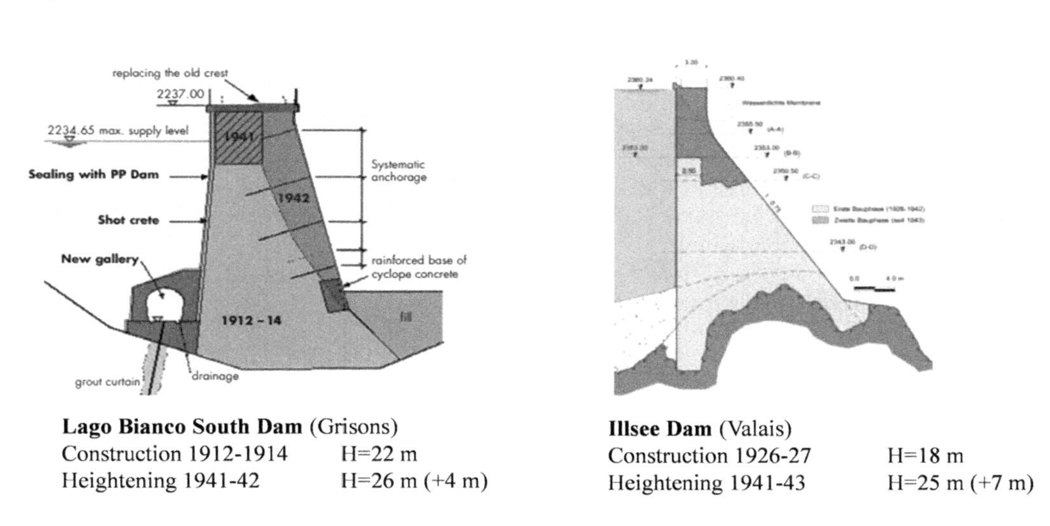


Figure 2. Lago Bianco South dam and Illsee dam.

Raised by 7 m in the early 1940s, Illsee dam, built in 1926-27, is shown in Figure 2, right. Like the other two dams described above, Illsee dam axis is curved in plan in its main section, where it crosses the talweg. Over the years, the structure has been found to suffer from AAR, leading to significant and irreversible deformation and cracking. In addition, the dam did not meet modern safety requirements, having been designed without regard to seismic loading. Considerable rehabilitation works were carried out in 2012-13 (concrete sawing to relax the bank-bank stresses caused by the RAG, installation of vertical prestressed anchors, and reconstruction of the dam crest).

Barcuns dam, built in 1947 and shown in Figure 3, is the gravity dam that was most recently raised in Switzerland, that is in 2013-14. Its 5 m heightening is remarkable in that it was carried out on its upstream face, unlike all the dams described above. This solution is constraining because it requires the reservoir to be emptied. It was possible to implement it at Barcuns because the entire hydraulic circuit and hydroelectric power station were rehabilitated simultaneously. The facility was shut down for around two years to carry out this work.

**Barcuns Dam** (Grisons)

| | |
|---|---|
| Construction 1947 | H=31.8 m |
| Heightening 2013-14 | H=36.8 m (+5 m) |

Figure 3. Barcuns dam, with view from upstream of the dam heightening works.

### 3.2 *Compliance with safety standards*

In Switzerland, two dams are known to have been raised to improve their safety. These are List and Muslen gravity dams, both of which were reinforced and raised in 1982. The respective cross-sections are shown in Figure 4 below. During the 1970s, the safety requirements for dams were updated and these two dams did not meet the overturning criterion, thus generating high tensile stresses on the upstream face. When they were designed in the beginning of the 20$^{th}$ century, uplift pressure was not considered in dam design, nor were exceptional load cases (floods, earthquakes). In addition, the spillways on the two dams had insufficient capacity.

Faced with the need for major reinforcement work, the owners of these two facilities took the opportunity to raise their respective dams, resulting in the concrete covering of the old dam, which is no longer visible or accessible.

### 3.3 *Early commissioning of the scheme*

The two dams presented in this section are special cases of heightening. These dams were raised immediately after the construction of the initial dam, which was considerably reduced in size. The aim of this two-stage construction was to enable the hydraulic circuit to be commissioned ahead of time and to start producing hydroelectric power, without having to wait for the final size of the dam to be built.

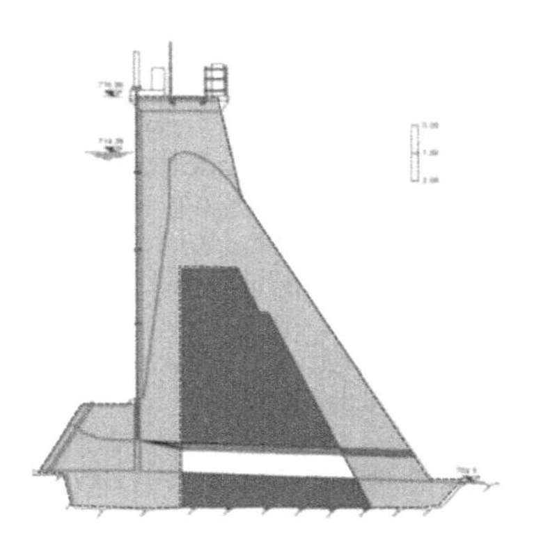

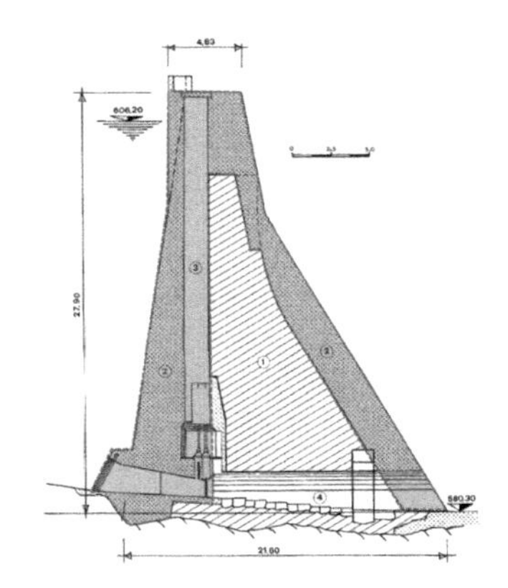

**List Dam** (Appenzell Rh.-Ext.)
Construction 1900-1901 H=13.5 m
Heightening 1982 H=16.5 m (+3 m)

**Muslen Dam** (St-Gall)
Construction 1909 H=23 m
Heightening 1982 H=28 m (+5 m)

Figure 4. List dam and Muslen dam.

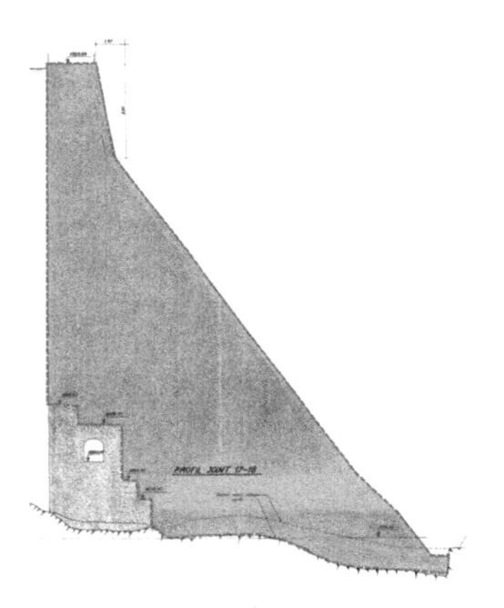

**Salanfe Dam** (Valais)
Construction 1947-1950 H=14.5 m
Heightening 1951-1953 H=52 m (+37.5 m)

Figure 5. Salanfe dam. On the right, photo of the first stage dam, H=14.5 m.

Salanfe dam is the first example shown in Figure 5 above. The first stage of the dam was built in 1947-50, with the second stage following in 1951-53. Miéville hydroelectric power plant, almost 1500 m lower down on the Rhône Valley, came into service in 1951.

This dam also suffered from AAR, which led to concrete sawing work on the structure in 2012-13 to release the compressive stresses generated by the reaction. The dam crest was also refurbished during this work.

The second case discussed in this section is that of les Toules arch dam, illustrated in Figure 6 below. Construction in two stages enabled Pallazuit power station to be commissioned in 1958 but was also made necessary by the road to the Grand-St-Bernard pass, which runs along the right bank of the valley and had to be raised some fifty metres before the reservoir could be impounded.

It can be seen that the first structure is a single-curvature arch dam, while the second is a double-curvature arch dam. The joint between both structures proved particularly difficult to manage. Added to this was the very wide shape of the valley, which is not very suitable for an arch dam (ratio of crest length to height > 5) and geological conditions which are not very homogeneous, particularly on the left bank, it was necessary to reinforce this dam by adding concrete strengthening on both downstream banks. This work took place in years 2008-11.

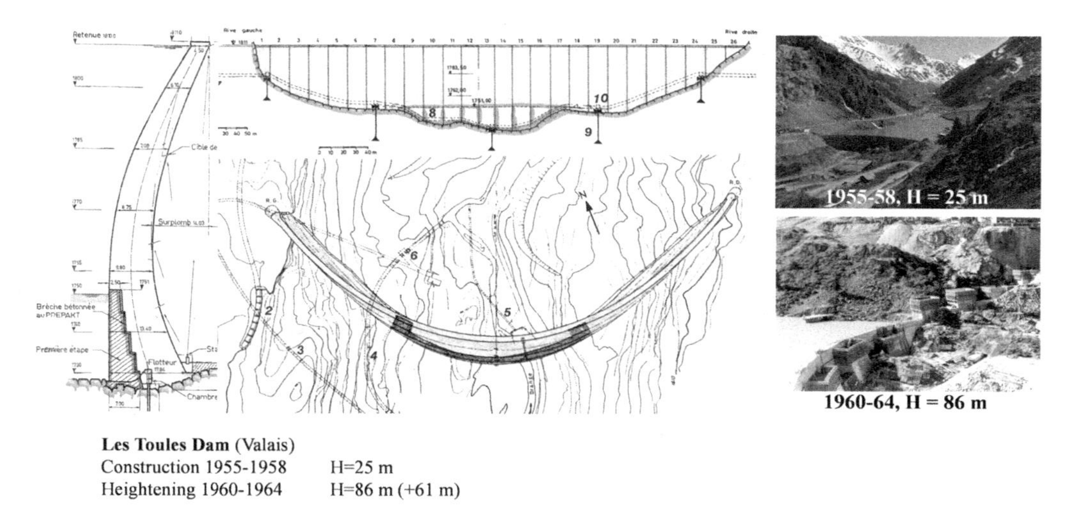


Figure 6. Les Toules dam. On the right, photos of the two stages of construction. It should be noted that this dam was further reinforced between 2008 and 2011 on its downstream face, using around 60000 m$^3$ of concrete. This is not shown on the figure above.

## 4 RECENT DAM HEIGHTENING

Three of Switzerland's major arch dams have been raised since the 1990s. They are Mauvoisin, Luzzone and Vieux Emosson dams.

There are only two ways of raising an arch dam:

- Either by propagating the shape of the upstream and downstream faces upwards. This is naturally the most natural and simple method. However, it is not always feasible, depending on the shape of the two faces. Mauvoisin and Luzzone dams were raised using this method.
- Or by modifying the shape of the arch. This method is considerably more complicated and costly than the previous one, as it involves partially deconstructing the existing arch dam before raising it. This is the case with Vieux Emosson dam.

### 4.1 *Mauvoisin dam*

The purpose of raising Mauvoisin dam is to transfer energy from summer to winter. Energy production in summer exceeds demand, while energy production in winter is insufficient to cover the country's needs. Thanks to the favourable hydrological conditions in the Haut Val de Bagnes, the operator (Forces Motrices de Mauvoisin) had the opportunity to increase winter energy production by 100 million KWh while reducing summer production by the same amount. This could be achieved by increasing the storage capacity of Mauvoisin reservoir by around 30 million m$^3$.

With a height of 236.5 m, Mauvoisin arch dam was the highest dam in operation in Europe in 1989. It had a crest length of 520 m and a crest width of 14 m at 1962.5 m asl. The dam was raised by constructing an additional arch in the extension of the original dam faces. The width of the raised dam crest at an altitude of 1976 m asl is 12 m, incl. a 6.6 m wide road. A gallery 5 m wide and 5.5 m high, located in the raised section, provides access to the underground hydraulic power station located upstream of Mauvoisin scheme. The volume of concrete used is 80 000 m$^3$, representing 4% of the volume of concrete used in the original dam.

The heightening has a crest length of 540 m and consists of 28 blocks that are concreted between the vertical joints of the dam. Each block, approximately 18 m long and 13.5 m high, is concreted in 5 lifts of 2.7 m each. The volume of each lift varies between 400 m$^3$ and 650 m$^3$ of concrete.

Ancillary work on the surface spillway, the mid-level outlet gate and the surge tank near Fionnay power station was carried out in parallel with the main work on the dam.

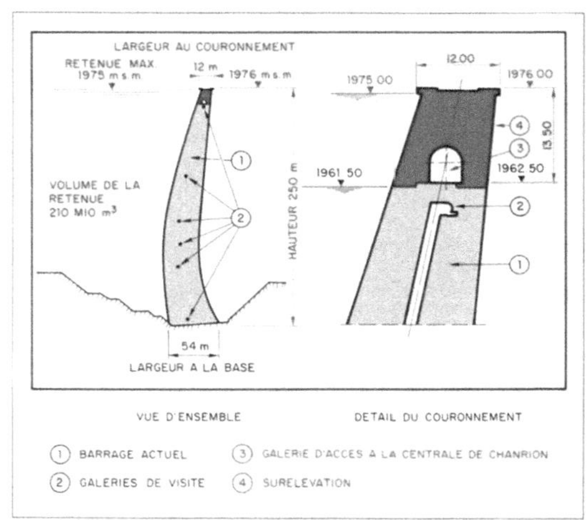

**Mauvoisin Dam** (Valais)
Construction 1951-1957 H=236.5 m
Heightening 1989-1991 H=250 m (+13.5 m)

Figure 7. Mauvoisin dam. Right, picture of the heightening job site (1990).

Work was carried out over three summer periods (April to October) from 1989 to 1991. Site installations began in April 1989. The batching plant, cement silos and crushing plant for preparing concrete aggregates were located at the foot of the dam, close to the large stockpile of materials left over from the construction of the dam. The cement was transported by train to Le Châble, and then by truck to the silos.

Concreting works were carried out from September 1989 to August 1991, using one crane (capacity 150 mt) in the first year and two cranes placed on the blocks already built, in 1990. By spring 1991, all that remained to be done was the concreting of the new dam crest. The joints between dam blocks were grouted between the end of May and the beginning of July 1991.

The raised dam was impounded in summer 1991, reaching its maximum level of 1975 m asl at the end of September. The dam behaviour was as expected by the design engineers.

### 4.2 *Luzzone dam*

Luzzone arch dam was built between 1959 and 1963. With a height of 208 m, a crest length of 550 m, a thickness of 36 m at the base and 10 m at the crest, the volume of concrete for this dam amounted to 1.32 million m$^3$ before heightening. The seasonal storage basin had a capacity of 87 million m$^3$ and is the main reservoir of the Blenio power plant. However, given the growing demand for energy in winter and the fact that the summer inflow of

127 million m$^3$ far exceeded the reservoir storage capacity, it became interesting to increase the storage volume.

A project to raise the dam by 17 m was designed and work began in 1995. Since 1999, the dam is 225 m high, has a concrete volume of 1.40 million m$^3$ and a storage capacity of 107 million m$^3$ (85% of average summer inflows), see Figure 8.

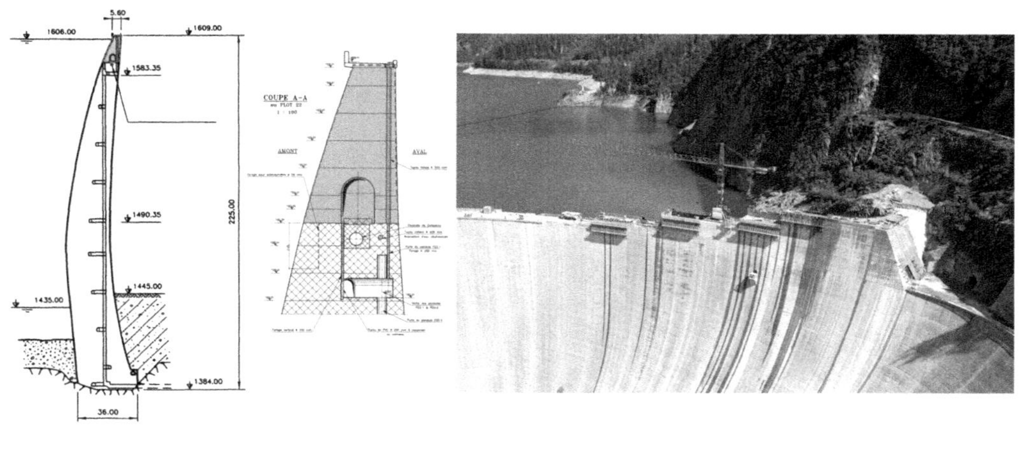


**Luzzone Dam** (Tessin)
Construction 1959-1963 H=208 m
Heightening 1995-1998 H=225 m (+17 m)

Figure 8. Luzzone dam. Right, picture of the heightening job site (1995), showing detail of the left bank abutment.

As shown above for Mauvoisin arch dam, the technical concept of the heightening involves the upward propagation of the geometric definition of the upstream and downstream faces. An additional arch was therefore added to the dam, bringing the width at the new crest to 5.6 m. The abutment on the left bank of the heightening is special. To save concrete, the raised section ends on an artificial abutment in the shape of a gravity dam (see Figure 8, photo right) and in order to have the compression forces of the arch diving down into the existing section, the bloc joints near the artificial abutment were only partially grouted. To maintain access to a mountain pasture upstream of Luzzone reservoir during and after the heightening work, a road gallery was integrated into the raised dam section.

Concreting was carried out using a crane placed halfway up the raising, which moved as the work progressed from the left bank to the right bank. As with the original dam, the aggregates came from a moraine quarrel located in a side valley a few kilometres away. The batching plant was located on the right bank of the dam and produced a total of 80 000 m$^3$ of dam concrete, with a binder content of 250 kg/m$^3$ and a maximum size aggregate of 63 mm. The binder contained 80% Portland cement and 20% fly ash, making it possible to give up artificial cooling of the concrete (post-cooling).

In parallel with the concrete work on the dam, the spillway and surge tank were raised and a new access gallery to the mountain pasture upstream of the reservoir was excavated.

The heightened dam has been in service since 1999, behaving satisfactorily and in line with its design.

### 4.3 *Vieux Emosson dam*

Vieux Emosson dam is located in Canton of Valais, close to the border with France. The raising of the dam is a key element of the vast Nant de Drance pumped-storage scheme, which came into operation in 2022. This dam serves as an upper reservoir, and the raising of the dam allowed more than doubling the active storage capacity (from 11.2 million $m^3$ to 24.6 million $m^3$), giving the pumped-storage scheme greater flexibility.

The dam was built in the 1950s by Swiss Federal Railways. The cross-section of the dam corresponds to a 62 000 $m^3$ concrete gravity dam, which was actually a single-curvature arch dam. The heightening consists of a double-curvature arch dam with horizontal and vertical sections formed by parabolic segments. Geometric constraints due to the shape of the valley and the first dam meant that the upper part of the latter had to be demolished and removed (17 000 $m^3$ of concrete), before the raising could be built (65 000 $m^3$ of concrete) and the transition made from the initial single-curvature to the double-curvature of the raising. The total volume of concrete for the raised structure amounts to 110 000 $m^3$. The heightening concept is shown below in Figure 9.

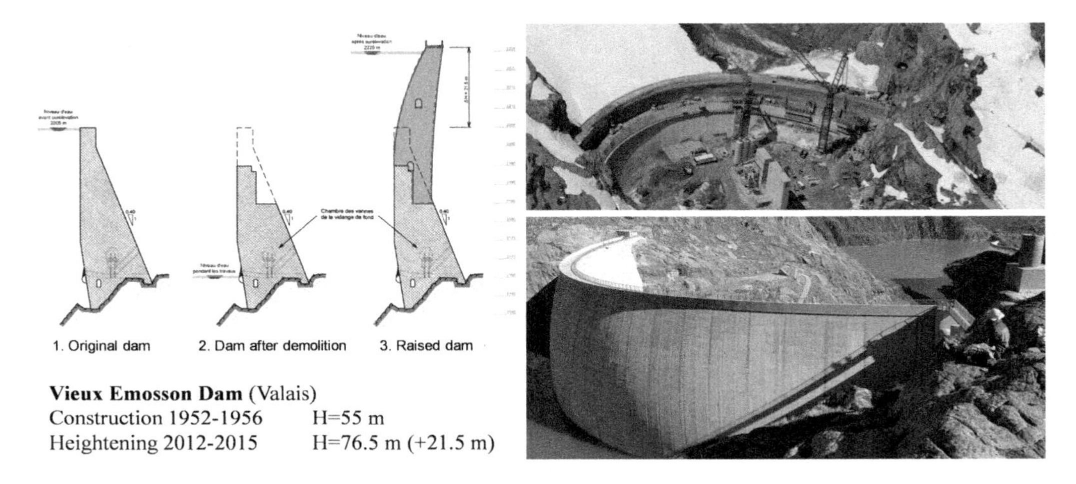


Figure 9. Vieux Emosson dam. Right, photos of the heightening work.

On the right bank, the first dam is just above the surface of the rock mass. On this bank, the crest of the raised dam is clearly above the rock mass. For this reason, a wing wall had to be built at the end of the arch. Its function is to close off the reservoir and house the spillway, which is a free flow ungated spillway.

The reservoir was empty from the start of the demolition work until the raising of the vault was completed. During this period, the spillway gates remained open and were being refurbished. However, at least one valve always remained operational so that the flow could be managed in the event of flooding.

The demolition of the upper part of the first dam was a critical point and a major challenge for the project. Initially, blasting was planned to demolish the dam concrete over a height of around 20 m. However, this technology proved unsuitable. The demolished surfaces had discontinuous and rugged shapes and the allowable shaking limits were difficult to comply with. Finally, the efficiency of demolition by blasting was low. Following this observation, the contractor gave up the method and switched to demolition of the concrete by using road-type machines (planer machines) that travelled back and forth over the crest. Milling machines were used to demolish the concrete that the large planers could not reach. This demolition method proved to be far more efficient than blasting, while limiting the problem of vibrations generated by blasting. The demolition work lasted from mid-June to the end of October 2012.

Concreting took place over two summer seasons (April to October), in 2013 and 2014. The concrete aggregates were produced lower in the valley and transported by truck to the toe of the dam, where the batching plant was located. The concrete was transported onto the dam using two tower cranes, also installed downstream of the dam.

In spring 2015, the dam blocs were grouted in two stages over the raised height. The dam was then commissioned over the following years to the satisfaction of the design engineer and the supervisory authorities.

## 5 DAM HEIGHTENING ABROAD

Swiss engineering know-how in the field of (large) dams is widely recognised throughout the world and has historically been exported well throughout the 20th century and into the early 21st century. There are many examples, and some of the most daring projects, to be found in numerous references and publications on the matter.

The same is certainly true for the very specific field of dam heightening. Although this type of work is more recent and its applications still rarer, Swiss engineering firms have had a few opportunities to demonstrate their capacity for innovation and their technical expertise in various countries. Some outstanding examples are described briefly below.

In the early 2000s, the raising of the Ekbatan buttress dam in Iran, built between 1959 and 1963, was designed and dimensioned by Swiss engineers. The project involved raising the dam by 25 m, from 54 m to 79 m. The project was built and successfully commissioned in the early 2010s.

In Angola, the Cambambe double-curvature arch dam on the Kwanza River, 180 km east of the capital Luanda, was raised between 2012 and 2018 (Figure 10). The design, execution project and supervision of the works were carried out by Swiss engineering firms. Particularly slender, and located in a spectacular natural landscape, the heightening project has given rise to a number of publications. In particular, the construction of the elevation had to be planned according to the flow of the river, continuously spilling over the existing dam.

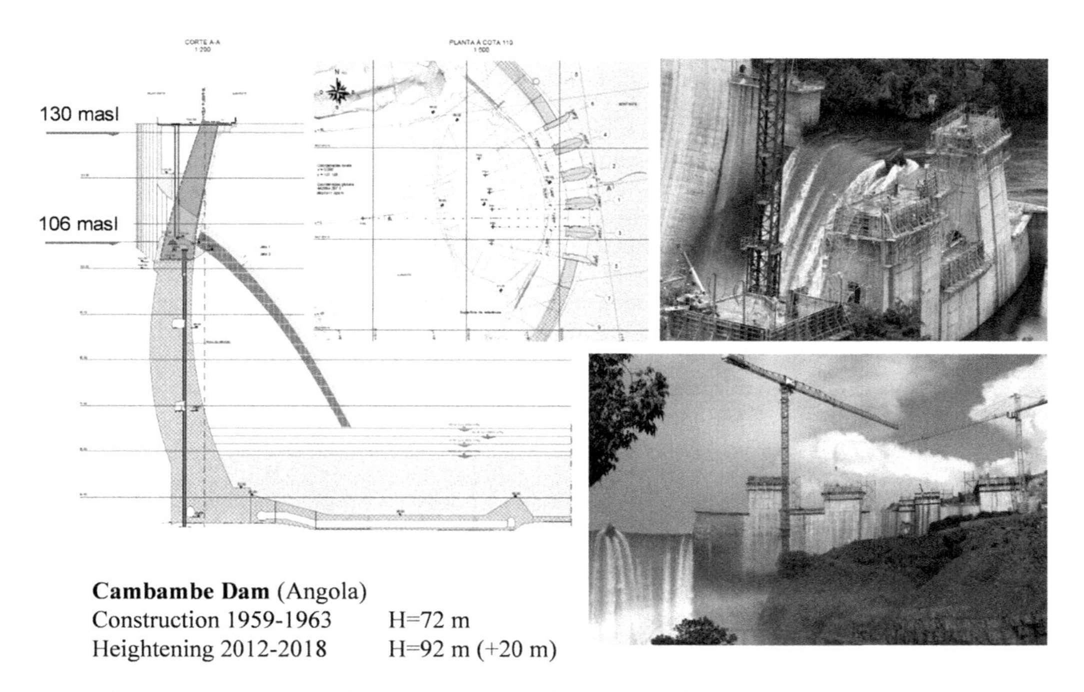


Figure 10. Cambambe dam. Right, photos of the heightening works.

Finally, there is the Limberg arch dam in Austria, near Kaprun/Zell am See, whose initial height of 120 m will be raised to 129 m. Studies are ongoing and work is scheduled to begin in the coming years. A Swiss engineering firm is in charge of the design and production of the construction project.

## 6 CONCLUSIONS AND FUTURE PROSPECTS

### 6.1 *Retrospective*

Dams have been raised in Switzerland for over a century. Twelve examples have been listed and described in this paper. The size of the structures raised varies greatly, ranging from around ten metres for the most modest to over 200 m high for Luzzone and Mauvoisin arch dams.

The many advantages of such challenging projects have been highlighted, both from an environmental and an economic point of view. Generally speaking, the gain of a relatively large additional water storage volume compared with the often moderate scale of the heightening work and its impact on the environment and landscape makes raising a dam economically attractive.

The review showed that, in Switzerland, the majority of projects involve raising the height of concrete dams; however, there are also heightening projects for embankment dams. Work to raise Plans Mayens embankment dam in Crans-Montana is scheduled for 2023-2025.

Given the diversity of each dam in its natural environment and the unique nature of each facility, each raising project requires the implementation of specific solutions tailored to the site and the problems encountered, as it has been amply demonstrated throughout this retrospective. The range of technical solutions deployed over the last century demonstrates the expertise of Swiss engineering in this field, which has also been exported.

### 6.2 *Future prospects*

In the current geopolitical and energy context (growing population, rising electricity consumption, political determination to move away from nuclear power, heavy dependence on fossil fuels such as gas and oil, unstable political situation in Europe, climate crisis, strong development of solar and wind energy), the need to create energy storage facilities for the winter is becoming increasingly obvious. Only hydroelectric schemes with their reservoir dams can offer the possibility of transferring water reserves (i.e. energy) from summer to winter, with the reservoirs acting as huge rechargeable batteries.

In this context and under the leadership of the Swiss government, the various stakeholders in the field of hydropower (cantons, universities, environmental and landscape protection associations, electricity companies) have come together in 2020-2021 in a round table to develop a common approach to the challenges facing hydropower in the context of the 2050 energy strategy, the zero-emissions climate target, security of energy supply and the preservation of biodiversity. The round table identified 15 hydroelectric storage power plant projects that, based on current knowledge, are the most promising in terms of energy production and whose implementation would have the least impact on biodiversity and the landscape. Their implementation would make it possible to achieve a cumulative additional adjustable winter production of 2 TWh by 2040. The list of these 15 projects is indicative and not exhaustive. The projects are eligible for investment grants of up to 60%.

Of the 15 projects highlighted by the round table in December 2021, at least 10 involve heightening of existing dams. The context is therefore very favourable for several dams to be raised in the coming years and decades, and numerous studies are under way in Switzerland.

### 6.3 *Final remark*

There is no official directory of heightened dams in Switzerland. As far as possible, this paper aims to present an exhaustive list of raised dams in this country. However, it cannot be ruled out that some dams may have escaped the authors' attention.

This research would not have been possible without the active collaboration of several operators (Alpiq, Axpo, Groupe E, KWO) and engineering firms (Afry, Gruner Stucky, Lombardi), as well as the supervisory authority (SFOE) and the ETHZ, all of whom contributed to this work by sharing the information and data in their possession. Our sincere thanks to all of them.

## REFERENCES


Amman E. 1985. La transformation des barrages de Muslen et de List. *Commission Internationale des Grands Barrages, Quinzième Congrès des Grands Barrages, Lausanne*: Q59, R23, 381-394.

Dams in Switzerland, Source for Worldwide Swiss Dam Engineering. 2011. *Swiss Committee on Dams.* Baden-Dättwil: buag

Felix, D., Müller-Hagmann, M. & Boes, R. 2020. Ausbaupotenzial der bestehenden Speicherseen in der Schweiz. *Wasser, Energie, Luft – Eau, Energie, Air, Heft 1*: 1–10

Feuz, B. & Schenk, Th. 1992. Die Erhöhung der Staumauer Mauvoisin. *Wasser, Energie, Luft – Eau, Energie, Air, Heft 10*: 245–248

Feuz, B. 1994. Raising of the Mauvoisin Dam. *Structural Engineering International, International Association for Bridge and Structural Engineering, 2/94*: 103–104

Golliard, D., Lazaro, P., Demont, J.-B. & Favez, B. 2001. Travaux de réhabilitation de l'aménagement de l'Oelberg-Maigrauge, Fribourg. *Wasser, Energie, Luft – Eau, Energie, Air, Heft 3/4*: 63–70

Kressig, D. 2012. Realisierung Sanierung Talsperre Illsee. *Schweizerischer Wasserwirtschaftsverband, Fachtagung Wasserkraft*, Luzern.

Leite Ribeiro, M., Vallotton, O. & Wohnlich, A. 2022. Two Recent Cases of Arch Dam Raising, Lessons Learnt and Innovation. *Commission Internationale des Grands Barrages, Vingt-Septième Congrès des Grands Barrages, Marseille*: Q104c

Vallotton, O. 2012. Surélévation du barrage de Vieux Emosson. *Wasser, Energie, Luft – Eau, Energie, Air, Heft 3/4*: 209–215

Vallotton, O. 2015. Surélévation du barrage de Vieux Emosson. *Commission Internationale des Grands Barrages, Vingt-Cinquième Congrès des Grands Barrages, Stavanger*: Q99

Wohnlich, A. 2012. Surélévation du barrage-voûte de Cambambe, Angola. *Wasser, Energie, Luft – Eau, Energie, Air, Heft 3/4*: 216–219

Wohnlich, A. 2021. Barrage des Toules – Faiblesses structurelles, confortement du barrage. *Comité Suisse des Barrages, Journées d'études CSB*, Crans-Montana.

*Role of Dams and Reservoirs in a Successful Energy Transition – Boes, Droz & Leroy (Eds)*


# Swiss dam engineering in the world

Patrice Droz
*Swiss Committee on Dams*

ABSTRACT: Swiss dam engineers and contractors have always demonstrated a great interest in exporting their know-how and expertise in the domain of dams and hydropower. The article presents an overview of the most recent international projects.

RÉSUMÉ: Les ingénieurs et entreprises suisses ont depuis longtemps montré un intérêt à exporter leur savoir-faire et leur expérience à l'étranger dans le domaine des barrages et des aménagements hydroélectriques. Cet article présente un tour d'horizon des activités récentes hors de Suisse dans le domaine.

## 1 INTRODUCTION

The development of dams and hydropower in Switzerland started at the end of the 19th century, taking advantage of the topography of the country as well as its water resources. The development of hydropower encouraged the industrialization of the country as well as its electric railway system. But progressively, with their know-how gained in the Alps, Swiss engineers and contractors exported their experience abroad, firstly to Europe and then worldwide.

As the Swiss Committee on Dams is celebrating its 75th anniversary, it should be of interest to briefly describe some of the recent international projects in which Swiss engineering consulting firms and experts, contractors and research laboratories have been involved. Of course, an exhaustive treatment is not the aim of the present paper: a selection has been made by the author, taking into account the importance of the project, the difficulties encountered in the completion of the projects, and the specific techniques used, and solutions selected.

## 2 DEVELOPMENT OF NEW DAM PROJECTS

### 2.1 *Europe*

The 690 MW capacity, 600 m gross head Kárahnjúkar hydroelectric scheme (Iceland), harnesses water from two glacial rivers originating in the large Vatnajökull Glacier, stored behind the 198 m high Kárahnjúkar concrete-faced rockfill dam (CFRD). Difficult tunnel conditions due to large groundwater inflows and to locally unstable rock conditions were encountered during the excavation of the headrace tunnel. The resulting delays required acceleration of the filling, pressurizing and commissioning of the headrace tunnel. This was achieved by filling the headrace tunnel in two stages. First the penstocks and the lower section of the tunnel were filled using the groundwater inflow, and water stored in the lower tunnel behind a temporary cofferdam in the tunnel was used to commence the wet testing of several turbine units while finishing work continued in the upper section of the tunnel. Following completion of the tunnel finishing work, the remainder of the tunnel was filled and pressurized in a second stage (Kaelin 2009). The dam body is also located on a fault, which required the construction of a joint in the toe wall and the concrete face. Model studies on that project were conducted at both hydraulic laboratories of the Swiss Federal Institutes of Technology in Lausanne (Bollaert 2003) and Zurich (Berchtold and Pfister 2011).

The Neikotski dam is under construction in Northern Bulgaria, it is part of a water supply project. The maximum height of the dam is 47.2 m; the crest is 200 m long, and the embankment volume is approx. 300,000 $m^3$. The asphalt core rockfill dam (ACRD) is founded on rock. The asphalt-concrete core is vertical, located in the central section of the dam. It starts from the top of a grouting gallery which is embedded in the rock foundation. During the design phase, a comparison was made of seismic horizontal displacements obtained from a dynamic analysis and by two simplified analytic methods. In this case the results of the dynamic analysis and the simplified analysis were almost the same. However, the dynamic analysis using the finite element method provides much more information on the seismic behaviour of the dam (Tzenkov 2023).

Devoll Hydropower Project is located about 70 km southeast of the Albanian capital Tirana and consists of two hydropower plants, Banja and Moglice. The hydropower plants will have a total installed capacity of 256 MW and a mean annual production of about 703 GWh, increasing the Albanian electricity production by about 17 per cent. The lower one (Banja) was commissioned in 2016. The construction of the Moglice HPP began in 2014 with an installed capacity of 184 MW. The 167 m high dam is an ACRD with a crest length of 370 m and is one of the highest of its type in the world (Tirunas 2018).

## 2.2 *Middle East*

The recent development of hydropower projects and the construction of numerous large dams in Turkey offered the opportunity to Swiss consulting firms to be part of the design and construction of a couple of very significant dams in Anatolia.

The Deriner dam is a double curvature arch dam located on the Çoruh River in Northeastern Turkey. With a height of 249 m and a concrete volume of 3.5 $Mm^3$, it is currently the second highest dam in Turkey. The installed capacity of the powerhouse is 670 MW, with four Francis turbines, and its annual power generation amounts to 2118 GWh, accounting for approximately 1.1% of the total energy production in Turkey and roughly 6% of Turkey's hydropower generation capacity. State Hydraulics Works (DSI) of Turkey owns the project. Swiss consulting firms were part of the project either on the owner side (Wieland et al. 2008) or the Turkish contractor side (Müller 2009). The construction works started in 2000 and the dam was commissioned in 2012. The flood evacuation system is particularly impressive with two gated overflow spillways at the left and right abutments of the dam with a total capacity of 2,225 $m^3/s$ and eight mid-level bottom outlets with a total discharge capacity of 7,000 $m^3/s$. The total volume of excavated material amounted to 8.7 $Mm^3$, and for the stabilization of the abutments more than 2,000 2 MN post-tensioned anchors were installed (Figure 1).

Figure 1. Upstream view of Deriner dam and rock stabilization of abutments.

The Ilisu Dam & Hydroelectric Power Plant (1,200 MW) is located on the Tigris River in the Southeast of Turkey close to the border with Iraq (Figure 2). The power plant is composed of a 135 m high and 2,289 m long CFRD, a gravity dam section, a spillway, power intakes and power tunnels with a maximum diameter of 13 m, a powerhouse (6 Francis turbines with an installed capacity of 1200 MW), a tailrace channel and three river diversion tunnels (Ø12 m), one of them acting as bottom outlet. The construction was finished in 2020 (IM 2023, Stucky Gruner 2023).

Figure 2. Ilisu dam, powerhouse and spillway.

### 2.3 *Central Asia*

Rogun dam is under construction in Tajikistan and will be completed in stages until 2030. The dam is a 325 m high earth core rockfill dam (ECRD). Extensive studies were carried out in order to assess its impacts on the riparian countries in terms of water resources (Pöyry 2014). At present several Swiss consulting firms as well as contractors are taking an active part in the detailed design and the construction of the dam. The dam is located in a highly seismic region in the Pamir Mountains and an active fault passes through the footprint of the dam parallel to the dam axis. Further challenges are the difficult geological conditions with rock salt formations and the high sediment content of the Vakhsh River.

During the construction of the Sangtuda 2 HPP in Tajikistan, severe water inflows occurred within the excavation of the run-of-river type powerplant. A specific support was provided for understanding the particular hydrogeological conditions, which led to the drainage of a regional karst aquifer, and for designing appropriate mitigation measures (Ghader et Bussard 2013). A pumping of around 10 m3/s allowed the successful completion of the project.

Rudbar Lorestan is an earth core rockfill dam with a height of 156 m. It is located in a narrow canyon in the seismically very active Zagros Mountain Range in the west of Iran. The dam is subjected to multiple seismic hazards including ground shaking, movements along multi-directional discontinuities and faults in the dam footprint. Due to the presence of these discontinuities and secondary faults, the originally planned concrete gravity dam with a slip joint across the main fault was replaced by a conservatively designed earth core rockfill dam. The worst-case earthquake scenario is a magnitude 7.5 earthquake at a distance of 1.5 km from the dam site causing a horizontal peak acceleration of 0.75 g and maximal movements along faults in the dam footprint of 1.5 m. A large freeboard was provided to cope with seismic deformations of the dam body and the run-up of impulse waves caused by mass movements into the reservoir. The first reservoir impounding started in 2017 (Wieland 2019).

### 2.4 *Southeast Asia*

The Xayaburi Hydroelectric Power Project (Figure 3) is a run-of-river hydropower plant located in the mainstream of the Mekong River, in Lao PDR, approximately 100 km downstream of the city of Luang Prabang. The scheme includes a navigation lock, spillway and intermediate block, and the main powerhouse with an installed capacity of 1,285 MW.

The project also comprises state-of-the-art fish passing facilities for upstream and downstream migration. The project was commissioned in 2019 (Morier-Genoud 2019).

At present the 1460 MW Luang Prabang run-of-river power plant, located about 30 km upstream of Luang Prabang in Laos, is under construction in which Swiss dam consultants are involved.

Figure 3. View of Xayaburi dam.

Nam Ou VI, located along a tributary of the Mekong River in Laos, is an 88 m high rockfill dam with an upstream geomembrane. The composite PVC geomembrane, installed using an innovative design, is now increasingly adopted to construct rockfill dams, which allows the construction of embankment dams at lower costs. At Nam Ou VI the geomembrane system was installed in three stages (Scuero et al. 2016).

Son La and Lai Chau are RCC dams located in Vietnam. The highlight of those projects is the use of "pond ash" i.e. fly ash that went into a waste lagoon as a slurry as nobody in Vietnam had any use for such material. Son La was the first RCC dam globally that made use of such material as a major cementitious material in the mix. Lai Chau again used it, as well as other large RCC dams in Vietnam and later also in Laos. Both projects were completed ahead of schedule (Conrad et al. 2010, Conrad et al. 2014).

### 2.5 *Africa*

Worth mentioning are the design and construction supervision of a number of irrigation, water supply and flood protection dams in North Africa by various Swiss consulting firms. Swiss consultants also served as members of panels of experts for different dams in Africa and have advised on the formation of the Dam Safety Directorate in Ethiopia, the country with the largest number of dams under construction in Africa. In the past decade, a number of hydraulic model studies on African dam projects have been performed at the hydraulic laboratories of the Swiss Federal Institutes of Technology in Lausanne (Stojnic et al. 2018) and Zurich (e.g. Arnold et al. 2018). Water resources management studies are also worth to mention (Gamito de Saldanha Calado Matos and Schleiss 2017).

### 2.6 *Americas*

The Toachi-Pilatón hydroelectric power plant uses the water from the homonym Rivers, in the North-West of Ecuador. The power plant has an installed capacity of 255 MW and is connected to the 59 m high Toachi concrete gravity dam. Construction was completed in 2015 (Lombardi 2023).

The Cerro del Águila HPP is located in the Peruvian Andes about 270 km from the capital Lima. This new scheme on the Mantaro River includes an 88 m high RCC arch-gravity dam, a 5.7 km long headrace tunnel, a 242 m high pressure shaft, an underground powerhouse with an installed capacity of 510 MW, and a 1.9 km long tailrace tunnel. A gated crest spillway with a discharge capacity of 7,000 $m^3/s$ is integrated in the dam body as well as 6 bottom outlets with a total capacity of 5,000 $m^3/s$. At the dam toe, a 3 MW power plant makes use of the ecological flow. Construction was completed in 2016 (Lombardi 2023).

### 2.7 *Oceania*

It is worth mentioning that physical model tests of the power intakes and surge chamber system of the Snowy 2.0 pump-storage scheme, presently under construction in Australia, were carried out at the hydraulic laboratory of EPF Lausanne (PL-LCH EPFL 2023).

## 3 DAM REHABILITATION

### 3.1 *Aging*

The Enguri dam (Figure 4), located in the western part of Georgia, is a 272 m high arch dam and until recently was the highest arch dam in the world. The dam was completed in 1984, several years after the power production at reduced head started (the units were commissioned in 1978-80). Treatment of the dam foundation and other works continued until 1988, the first year in which the reservoir was allowed to be filled to its maximum level.

After independence from the former Soviet Union the dam suffered due to the lack of maintenance and political unrest. Among the major parts of the scheme affected were the dam hydro-mechanical works particularly the drainage system, electro-mechanical elements and the low-level outlets of the dam, all contributing to poor reliability and safety in operation of the scheme.

Figure 4. Enguri arch dam, Georgia.

An expert mission was then organized by a group of Swiss and French Engineers aimed at defining a rehabilitation program (Quigley et al. 2006), which consisted mainly of:

- General safety assessment and identification of the status of the monitoring equipment for the dam;
- Assessment of the feasibility of the rehabilitation of the entire project; and
- Definition of the scope of the work required for the dam body, pressure tunnel and powerhouse.

The first phase of the rehabilitation works extended until 2006.

Within the grid of Montenegro the Hydro Power Plant Piva plays an important role. It has been in operation since 1976 and needed to be rehabilitated. The plant is located in the north-western part of Montenegro, close to the border with Bosnia and Herzegovina. The high storage capacity of the basin guarantees a high plant utilisation factor, even during dry years. The 220 m high arch dam, also known as Mratinje Dam, is one of the highest arch dams in Europe. The hydro-mechanical and electrical-equipment as well as the civil structures, including the dam, intake and outlet structure needed rehabilitation and modernisation in order to extend their economic lifespan (Obermoser 2009). Based on a detailed investigation, short-term rehabilitation measures related to the dam were identified, including rock support in unlined galleries, improvement of the dam grout curtain and modernization of the dam monitoring system. The observed damage to the dam concrete were minor and could be addressed as part of regular maintenance.

The Studena dam in Bulgaria is a 55 m high buttress dam in a seismically active region. The dam, composed of 25 blocks, is used for water supply, hydropower generation, and flood protection. Heavy deterioration of the concrete face required complete rehabilitation. As the water supply could not be interrupted, rehabilitation works had to be performed, mostly underwater. A new watertight synthetic facing covers the upstream face of the dam. A major challenge was the repair of the upstream face because of its complicated geometry with complex intersecting concave corners requiring special fixation of the membrane and the very low temperatures during the construction work (Scuero et al. 2019).

The Kariba Dam is a 128 m high arch dam that was constructed between 1955 and 1959 across the Zambezi River that borders Zambia and Zimbabwe. The six centrally located submerged sluices form the spillway with a combined discharge capacity of 9,000 $m^3/s$. Prolonged spillages with a total volume of 511.1 $km^3$ through the floodgates from January 1962 to June 1981 resulted in an 80 m deep scour hole in the plunge pool immediately downstream of the dam (Figure 5).

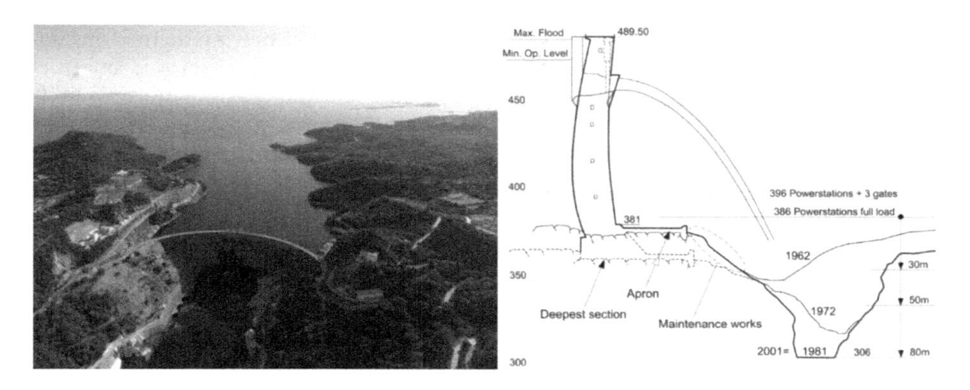


Figure 5. Kariba arch dam and the evolution of scouring close to the dam toe.

Numerical modelling and hydraulic model tests were carried out in order to define the geometry of the reshaping of the plunge pool to avoid further scouring (Stojnic et al. 2018). The remedial works are presently in progress (Mellal et al. 2023). In addition, the hydromechanical equipment of the 6 spillways, threatened by the development of alkali-aggregate reaction in the dam concrete, is being refurbished.

### 3.2 *Alkali-aggregate reaction*

The Chambon dam in France is an excellent example of how geomembrane systems can contribute to the extension of the lifespan of a dam. This 137 m high concrete gravity dam, completed in 1935, is affected by alkali-aggregate reaction (AAR). A series of slot cuttings were made and a drained exposed PVC geomembrane system was installed in 1994 to provide waterproofing protection at the upstream face. In 2013, the owner decided to carry out new slot cutting, and to strengthen the dam by means of an upstream system of tendons and carbon bands, which required removing the geomembrane system. The same geomembrane system was installed again. Rehabilitation works were successfully completed in 2014 (Scuero et al. 2016).

The Pian Telessio dam is an arch gravity dam completed in 1955 in Northern Italy. With a height of 80 m and a crest length of 515 m the dam impounds a reservoir with a capacity of 24 $Mm^3$. The dam thickness ranges from 5.7 m at the crest to a maximum of 35 m at its base. After approximately 20 years of operation the dam showed an upstream drift of up to 60 mm at the central pendulum in 2008. It was concluded that the permanent displacements were caused by the concrete expansion due to AAR. Rehabilitation works required the cutting of 16 vertical slots with a depth of 21 to 39 m, using a diamond wire saw (Amberg et al. 2009). Further analyses are still performed in order to assess the time-dependent decrease of the dam safety (Stucchi et al. 2023).

The Kainji dam is located on the river Niger in Nigeria. The plant suffers from AAR, especially the spillway structure, since a few years after its commissioning in 1968. The mass concrete structures on both sides of the spillway apply a thrust on the spillway structure. The mechanism is in general confirmed by monitoring results and visual observations. To mitigate the negative effect of AAR, rehabilitation works were carried out in 1996/97, comprising slot cutting for relief of compressive stresses, drilling of drainage holes and installation of additional monitoring instruments. Thanks to the previous rehabilitation works, there is sufficient evidence to conclude that the spillway can safely be operated within the coming years. Rehabilitation and upgrading of the monitoring instruments and improvement of the surveillance procedures are on the way (Ehlers et al. 2023).

### 3.3 *Seismic resistance*

The Fontanaluccia dam is a 40 m high and almost 100-year-old structure constituted by a central multiple-arch masonry dam body with the spillway and two side cyclopean concrete gravity dam sections. The dam is located in a narrow valley in the Italian Apennine region, a moderate-to-high seismic area. Following the release of new guidelines on the seismic safety assessment of dams in Italy in 2019, a re-assessment of the seismic safety of the dam was carried out. The results of the dynamic analyses showed that the multiple arch section is vulnerable to the earthquake action in the cross-canyon direction, which required seismic strengthening of the dam. A retrofit program was developed to enable the dam to remain in operation during the remedial works. (Abati et al. 2023).

### 3.4 *Dam heightening*

The Cambambe arch dam on the Kuanza River in Angola, was built from 1959 to 1963. From its initial conception, the dam was planned to be heightened in a later stage. However, for different reasons these works did not start until 2010. The initial arch dam was 72 m high, with a crest length of 250 m. The heightening of the dam was planned to be 20 m leading to a final dam height of 92 m. The heightening works (Figure 6) were completed in 2020 and took place at the same time as the rehabilitation of the existing power plant and the construction of a new open-air power plant. The management of the flood release during the construction of the dam-heightening required proper timing of the works (Wohnlich et al. 2012).

### 3.5 *Sediment bypass tunnels to counter reservoir sedimentation*

Nanhua reservoir in Taiwan suffers from large and ongoing sedimentation, threatening its sustainable operation. A large sediment bypass tunnel to route sediment from turbidity currents past the dam was planned and built, which was cross-checked by experts from the hydraulic laboratory of ETH Zurich (Boes et al. 2018). Similarly, the refurbishment of an existing sediment bypass tunnel with granite pavers at the Mud Mountain dam in the US Pacific Northwest was profiting from Swiss research and development performed at ETH Zurich (Auel et al. 2018), gaining an American engineering award. Another sedimentation model study

Figure 6. Heightening of Cambambe dam in Angola.

with settling pond and bypass tunnel was conducted at the ETH Zurich laboratory for a new reservoirs dam in Pakistan (Beck et al. 2016, Boes et al. 2019).

## 4 DAM SAFETY

### 4.1 *Post-seismic inspections*

The 106 m high Sefid Rud buttress dam, located in the Alborz Mountains in Iran, was completed in 1962. The dam was designed to withstand a peak ground acceleration (PGA) of 0.25 g. The dam was damaged by the magnitude 7.4 Manjil earthquake of the 21 June 1990, during which the nearby cities of Manjil and Rudbar were destroyed. The horizontal component of the PGA was estimated as 0.7 g. The main shock was followed by several strong aftershocks with magnitudes up to 6.0. The top portion of the dam was damaged. A large crack along the horizontal lift joint about 18m below the dam crest was observed on the upstream face of the dam involving all buttresses. In one buttress a wedge was created on the downstream face by a system of horizontal and inclined cracks along construction joints, which was displaced by about 30 mm. Spalling of concrete along the vertical block joints were also noticed as well as leakage through some of the cracks. Detailed inspection (Wieland et al. 2003) led to rehabilitation works including epoxy grouting and post-tensioned rock anchors.

On January 12, 2010, a magnitude 7.0 earthquake caused extensive damage and loss of lives in Haiti. The Péligre dam located about 60 km from the epicenter, was inspected in March 2010 (Droz et al. 2010). In the absence of appropriate monitoring instrumentation, only a thorough visual inspection could be made. No signs of structural damage were visible. However, because of the relatively large epicentral distance the level of ground shaking at the dam was rather low.

### 4.2 *Dam monitoring instrumentation and survey*

The Inga 1 (Figure 7) and 2 hydropower schemes are located on the Congo River, approximately 150 km southwest of Kinshasa. Both dams are of the buttress type and have shown irreversible movements downstream for decades. Unfortunately, the monitoring instrumentation of the dams, which was installed at the time of the construction, in the 1970s for Inga 1 and in the 1980s for Inga 2, appears to be incomplete, in poor condition, obsolete or inadequate to follow up the evolution of the deformations, in particular due to AAR, and to assess the safety of the dams regularly. In view of these problems a project, financed by the World Bank, was implemented (Droz et al. 2019):

- To improve the quality of the surveillance of the structures by restoring and enhancing the monitoring system of the dams and of the hydro plants as well as installing an adequate geodetic network; and
- To carry out the necessary investigations and studies to determine the causes of the irreversible movements observed in most of the structures of both Inga 1 and 2.

Figure 7. Inga 1 dam, DR Congo.

The Sardar Sarovar Dam is a concrete gravity dam built on the Narmada River in India. The dam was built to provide water and electricity to four Indian states. The gravity dam is 136 m high with a crest length of 1300 m. Almost 400 sensors were installed around 1994. The rehabilitation project was organized in 5 steps. In 2021 (step 1), functional tests of the existing instruments were carried out. Then, the system was upgraded from manual to semi-automatic by installing multiplexer boxes. In 2022, after the partial rehabilitation of the existing shafts (step 2), 3 new 76 m long direct pendulums were installed (step 3) with an automatic acquisition system for measuring the horizontal deformations of the dam. The commissioning of the complete automatic plumblines is planned for 2023 (step 4). By switching from manual and semi-automatic to fully automatic monitoring (step 5), the dam owner benefits from a quicker and safer monitoring of the dam (Ballarin et al. 2023).

Many of the dams in Sri Lanka are ageing and suffer from various structural deficiencies and shortcomings in their operation and maintenance procedures. To overcome these inadequacies, the Dam Safety and Water Resources Project (DSWRPP) was initiated in August 2008, with the assistance of World Bank financing. Besides rehabilitation of 32 major dam structures that show signs of deterioration, improvement of the monitoring data acquisition and analysis was performed (Sorgenfrei et al. 2011).

The installation of a monitoring system on and around the Usoy dam remains an exceptional project. Lake Sarez in in the Pamir Mountains in Tajikistan was formed in 1911 when a massive earthquake triggered a rockslide that buried the village of Usoy under a 650 m high mass of rock and ice debris, which dammed the Murghab River. The resulting 60 km long lake containing over 17 $km^3$ of water was created by the Usoy landslide dam, the highest dam in the world. Owing to its huge mass, the level of knowledge of the Usoy dam (Figure 8) is reduced to hypothesis based on observations and analysis of a few parameters gathered in the past, in particular the seepage rate evolution through the natural embankment. In order to reduce the risk related to Lake Sarez, a modern monitoring system and an early warning system were installed in 2006. The monitoring system covers an unstable slope, located some 4 km upstream of the dam, which, if it would fail, would create impulse waves that could threaten the stability of the dam or, at least, modify its seepage regime (Droz et al. 2006 and 2008).

Figure 8. Usoy landslide dam storing Lake Sarez in Tajikistan.

### 4.3 *Emergency preparedness plans*

Besides a safe design and construction of high quality as well as appropriate maintenance, monitoring and surveillance, the third pillar on which lays the safety of dams is the preparedness in case of emergency which enables to cope with residual risks. Recently, Emergency Preparedness Plans (EPP) have been elaborated for 20 dams in Turkey.

At present, the EPPs of Rogun dam (under construction), of Nurek dam a 305 m high ECRD located directly downstream Rogun and completed in 1980, and of Kariba dam (both under rehabilitation), are under elaboration.

### 4.4 *Capacity building*

A dam safety project incorporating five dams along the Drin and Mat Rivers was carried out in the northern part of Albania. One of the components of the project was the refurbishment and upgrading of the monitoring instrumentation for all five dams. The goal was to implement a durable modern monitoring system. The new instruments will help to improve the long-term safety of the dams. The safety, however, can only be improved if the installations are properly maintained, the instruments are frequently read, and the values are immediately evaluated and regularly analysed by specialized engineers. Hence training was the most important component for the sustainability of this investment. The instrument and software installations were completed with a comprehensive training program for both dam wardens and specialized engineers in charge for the data analysis and reporting (Stahl et al., 2013).

The Dam Safety Enhancement Program (DaSEP) was set in 2010. DaSEP aimed at improving the dam safety procedures of the thousands of dams under the responsibility of the Chinese Ministry of Water Resources (Méan et al. 2012). The various missions of Swiss experts and the training of Chinese counterparts in China and in Switzerland resulted in the modification of the legal dam safety regulatory framework in China, including, for example, the obligation to prepare annual safety reports and define clearly roles and responsibilities for small and medium size dams.

In the wake of the failure of the Xepian-Xenamnoy saddle dam in Laos in 2018 (Schleiss et al. 2019), a nationwide dam safety inspection was launched. The synthesis report of the main findings of the various inspections pointed out the necessity to improve the dam safety organization and procedures in Lao PDR including technical (Droz et al. 2022) and institutional (Darbre et al. 2022) aspects. This Swiss Agency for Development and Cooperation project is in progress and has already given fruitful results with the membership of Laos as full member of the ICOLD community, the modification of the Dam Safety Law, and the creation of a department specifically in charge of dam safety.

## 5 CONCLUSIONS AND PERSPECTIVES

After a long period of dam construction in Switzerland, owners, consulting firms and contractors, like their European colleagues, are facing aging problems of these dams, which on average are 75 years old. Coping with the aging problem requires rehabilitation works and the development of new solutions. The know-how gained may be used in other projects worldwide, including new dam projects, dam safety assessment, dam rehabilitation, as well as dam heightening and capacity building. Nowadays, due to climate change issues and increasing energy demand, new projects are under development in Switzerland.

## ACKNOWLEDGMENTS

The preparation of this paper would not have been possible without the support of the following consulting firms, contractors, organisations and research institutes: AFRY, BG, Carpi, EPFL (PL-LCH), ETH Zurich (VAW), Gruner Stucky, Norbert Géologues, Helvetas, Huggenberger, IM Maggia, Lombardi, Rittmeyer and Walo.

## REFERENCES

Abati, A., Gatto Frezza, G. A., Sparnacci, R. 2023.Seismic safety assessment of an old multiple arch gravity dam. *Symposium "Management for Safe Dams" - 91st Annual ICOLD Meeting – Gothenburg.*

Amberg, F., Bremen, R., Brizzo, N. 2009. Rehabilitation of the Pian Telessio dam (IT) affected by AAR-reaction. *23rd ICOLD Congress, Brasilia, Q. 90.*

Arnold, R., Bezzi, A., Lais, A., Boes, R.M. (2018). Intake structure design of entirely steel-lined pressure conduits crossing an RCC dam. *Proc. Hydro 2018 Conference*, Gdansk, Poland: Paper 04.02. Aqua~Media International, Wallington, UK.

Auel, C., Thene, J.R., Carroll, J., Holmes, C., Boes, R.M. (2018). Rehabilitation of the Mud Mountain bypass tunnel invert. *Proc. 26th ICOLD Congress*, Vienna, Austria, Q.100-R.4: 51–71.

Ballarin, A., Gardenghi, R., 2023, Huggenberger Communication.

Beck, C., Lutz, N., Lais, A., Vetsch, D., Boes, R.B. (2016). Patrind Hydropower Project, Pakistan Physical model investigations on the optimization of the sediment management concept. *Proc. Hydro 2016 Conference*, Montreux, Switzerland: Paper 26.08.

Berchtold, Th., Pfister, M. (2011). Kárahnjúkar Dam spillway, Iceland: Swiss contribution to reduce dynamic plunge pool pressures generated by a high-velocity jet. in: *Dams in Switzerland.* Swiss Committee on Dams, pp. 315–320, ISBN 978-3-85545-158-6, Switzerland.

Boes, R.M.; Beck, C.; Lutz, N.; Lais, A.; Albayrak, I. (2017). Hydraulics of water, air-water and sediment flow in downstream-controlled sediment bypass tunnels. In *Proc. 2nd Intl. Workshop on Sediment Bypass Tunnels* (Sumi, T., ed.), paper FP11, Kyoto University, Kyoto, Japan.

Boes, R.M., Müller-Hagmann, M., Albayrak, I., Müller, B., Caspescha, L., Flepp, A., Jacobs, F., Auel, C. (2018). Sediment bypass tunnels: Swiss experience with bypass efficiency and abrasion-resistant invert materials. *Proc. 26th ICOLD Congress*, Vienna, Austria, Q.100-R.40: 625–638.

Bollaert, E., Tomasson, G. G., Gisiger J.-P., Schleiss, A. 2003. The Karahnjukar hydroelectric project: transient analysis of the waterways system. *Wasser Energie Luft, Heft 3, 2003.*

Conrad, M., Dunstan, M.R.H., Morris, D., Pham Viet An, 2010. The Son La RCC dam – Testing in-situ properties, *Hydro Asia 2010.*

Conrad, M., Morris, D., Nguyen Phan Hung 2014. The RCC dam for the Lai Chau HPP- RCC full scale trials and challenges in the construction of the RCC dam, *Hydro Asia 2014.*

Darbre, G., Droz, P., Malaykham B. 2022. Institutional organization for dam safety in Lao PDR, *ICOLD Q105/R1, June 2022*, Marseille.

Droz, P., Darbre, G., Malaykham, B. 2022. Emergency dam safety inspections in Lao PDR, *ICOLD Q105/R38, June 2022 , Marseille.*

Droz, P., Fumagali, A., Novali, F., Young, B. 2008. GPS and INSAR technologies: a joint approach for the safety of lake Sarez, *4th Canadian Conference on Geohazards, Université Laval, mai 2008.*

Droz, P., Hegetschweiler, D 2010. Inspection du barrage de Péligre suite au séisme du 12 janvier 2010 en Haïti, *Eau Energie Air ( WEL ), 3–2010.*

Droz, P., Spacic-Gril, L. 2006. Lake Sarez Risk Mitigation Project: a global risk analysis, *Q86 – R75, 22nd ICOLD Conference on Large Dams, 2006, Barcelona.*

Droz, P., Wohnlich, A. 2019. Rehabilitation of the monitoring system of Inga 1 and 2 dams, *Hydropower & Dams, Issue 2*, 2019.

Ehlers, S., Goltz, M., 2023. The development of Alkali Aggregate Reaction (AAR) at the Kainji spillway structure after 50 years of operation, *Hydro Africa, Uganda.*

Gamito de Saldanha Calado Matos, J. P., Schleiss, A. 2017. A free and state-of-the-art probabilistic flow forecasting tool designed for Africa. *Proceedings of Int. Conference Africa 2017, Marrakech, Morocco.*

Ghader, S., Bussard., T. 2013. The study of drainage and water pumping of spillway and hydropower plant foundation in Sangtuda 2 project. Tajikistan. *1st Iranian Conference on Geotechnical Engineering, 22–23 October 2013. University of Mohaghegh, Ardabil, Iran IM, 2023, web site.*

Kaelin, J., Olafsson, S., Leifsson, P.S. 2009. Filling and pressurizing the headrace tunnel at Karahnjukar in Iceland. *Hydro2009, Lyon, France.*

Lombardi, 2023, *web site.*

Méan, P., Droz P., Cai, Y., Sheng, J 2012. Dam Safety Enhancement Program: A Cooperation Project between Switzerland and China, *Int. Symposium on dams for a changing world, ICOLD, Kyoto.*

Mellal, A., Chibvura, C., Mhlanga, S. Z., Nkweendenda A, Munodawafa, M. C., Quigley, B.M., Arigoni, A. 2023. Use of computational modelling for prediction and follow up of dam behaviour during plunge pool reshaping of the Kariba Dam. *Symposium "Management for Safe Dams" - 91st Annual ICOLD Meeting – Gothenburg 13–14 June 2023.*

Morier-Genoud, G. 2019. Fish pass design on the Mekong River – Challenges and lessons learned from Xayaburi HPP, *SHF Conference HydroES 2019, Grenoble, France, January 2019.*

Obermoser, H., Pješčić, S., Conrad, M., 2009. Focused Rehabilitation of the Piva HPP in Montenegro, *Conference Hydro 2009, Lyon, France.*

PL-LCH EPFL 2023, *Personal communication.*

Pöyry 2014, https://www.worldbank.org/en/country/tajikistan/brief/final-reports-related-to-the-proposed-rogun-hpp.

Quigley B., Matcharadze G. 2006. Upgrading and Refurbishment of Enguri Dam and Hydro Power Plant in Georgia, *HYDRO 2006, Porto Carras, Greece.*

Schleiss, A., Tournier, J.-P., Chraibi, A. 2019. XPXN-Saddle Dam D failure - IEP Final Report, *ICOLD annual meeting, Ottawa, 2019.*

Scuero, A., Vaschetti, G., Machado do Vale, J. 2016. A unique case: new works at Chambon dam. *International Symposium on "Appropriate technology to ensure proper Development, Operation and Maintenance of Dams in Developing Countries", Johannesburg, South Africa.*

Scuero, A. Vaschetti, G. 2019. Underwater technologies for rehabilitation of dams: Studena case history. *ICOLD annual meeting, Ottawa, 2019.*

Scuero, A. Vaschetti, G. J. Cowland, J., Cai, B., Xuan, L. 2016. Nam Ou VI: Geomembrane Face Rockfill Dam in Laos. *Proceedings of Asia 2016, Vientiane, Lao PDR, 2016.*

Sorgenfrei, A., Friedrich, M. 2011. Dam Safety and Operational Efficiency Improvements in Sri Lanka, *ICOLD 70th Annual meeting, Lucerne, Switzerland,* June 2011.

Stahl, H., Celo, M. 2013. Refurbishment and Upgrading of the Monitoring Instrumentation for Fierza and Ulza Dams in Albania, *Proc. ICOLD European Club Symposium held in Venice, Italy, Paper B18.*

Stojnic, I., Ylla, C., Amini A., De Cesare, G., Bollaert, E.F.R., Mhlanga, S.Z., Mazidza, D.Z., Schleiss, A.J. 2018. Kariba plunge pool rehabilitation, *HYDRO 2018 proceedings, Gdansk, Poland.*

Stucchi, R., Catalano, E., 2023. Numerical modelling of the Pian Telessio dam affected by AAR. *Proc. 12th ICOLD European Club Symposium "Role of dams and reservoirs in a successful energy transition", Interlaken, Switzerland* (Boes, R.M., Droz, P. & Leroy, R., eds.), Taylor & Francis, London.

Stucky Gruner, 2023, *web site.*

Tirunas, D., Aspen, B.V. 2018. The Moglice Hydropower Project, Albania – Construction Design Experience. *Proc. of HYDRO 2018, Gdansk, 15.-17.10.2018. Int. J. on Hydropower and Dams.*

Tzenkov, A.D., Kisliakov, D.S., Schwager, M. 2023. An application of sophisticated FEM and simplified methods to the seismic response analysis of an asphalt-concrete core rockfill dam. *Proc. 12th ICOLD European Club Symposium "Role of dams and reservoirs in a successful energy transition", Interlaken, Switzerland* (Boes, R.M., Droz, P. & Leroy, R., eds.), Taylor & Francis, London.

Wieland, M., Aemmer, M. and Ruoss, R. 2008. Designs Aspects of Deriner Dam. *Int. Water Power & Dam Construction, Volume 60 Number 7, pp* 19–23.

Wieland, M., Brenner, R.P., Sommer, P. 2003. Earthquake resiliency of large concrete dams: damage, repair, and strengthening concepts, *ICOLD Conference, Montreal, 2003.*

Wieland, M., Roshanomid, H. 2019. Seismic design aspects and first reservoir impounding of Rudbar Lorestan rockfill dam. *Proc. Symposium on Sustainable and Safe Dams around the World, ICOLD 2019 Annual Meeting, Ottawa, Canada, June 9–14, 2019.*

*Role of Dams and Reservoirs in a Successful Energy Transition – Boes, Droz & Leroy (Eds)*
*© 2023 The Author(s), ISBN 978-1-032-57668-8*

# Dam surveillance in Switzerland: A constant development

Henri Pougatsch
*Civil engineer EPF, Formely Commisionner for Dam Safety Water and Geology*

Isabelle Fern
*Civil engineer EPFL, Head of hydraulic team, Lombardi SA Ingénieurs Conseils, Fribourg, Switzerland*

ABSTRACT: The main purpose of this article is to recall the essential elements that govern the installation of a monitoring system whose purpose is to follow-up the behavior of water retention structures. It describes the past, the present and the future of this type of installation.

RÉSUMÉ: Le propos principal de cet article est de rappeler les éléments essentiels qui régissent l'installation d'un dispositif d'auscultation dont le but est d'assurer le suivi du comportement des ouvrages d'accumulation. Il fait état du passé, du présent et du futur de ce type d'installation.

## 1 INTRODUCTION

In the 1920s, Switzerland undertook many constructions of large dams. Specific monitoring systems were developed to gather data of the dams and its surroundings to assess their behaviours, guarantee safety and improve the design of new ones. The Montsalvens dam, in Canton of Fri-bourg, with a height of H=55m, was the first to be equipped with monitoring instruments. Using triangulation and leveling (Figure 1), clinometers and thermometers at different points, it was possible to assess the deformations for different levels of restraint and thermal stresses. The purpose of these measurements was to confirm the correctness of the calculation assumptions. The Federal Office of Topography was responsible for the triangulation and leveling measurements. In 1932, a pendulum reading system, developed by Juillard, was installed at the Spitallamm dam (Canton of Bern). This instrument became an essential monitoring device for Swiss dams and around the world.

Gradually, the monitoring of dams became predominant and the requirements for this mode of monitoring evolved. It resulted in increased precision and simplification of measurements. In addition, the computer processing of the data allowed a powerful analysis. Given the importance attributed to the monitoring device, adaptations have been made periodically in relation to the knowledge acquired and the new requirements.

## 2 LEGAL REQUIREMENTS

With the intensification of the construction of dams, the Swiss authorities deemed useful to complete the legislation, which was limited at that time to article 3bis of the federal law on the water police of June 1877, which also applied to dams. When it was drafted, the overriding im-portance of dam monitoring was recognized. An article of this regulation specifies that installations adapted to the dimensions of the structure will be fitted with so that it is possible

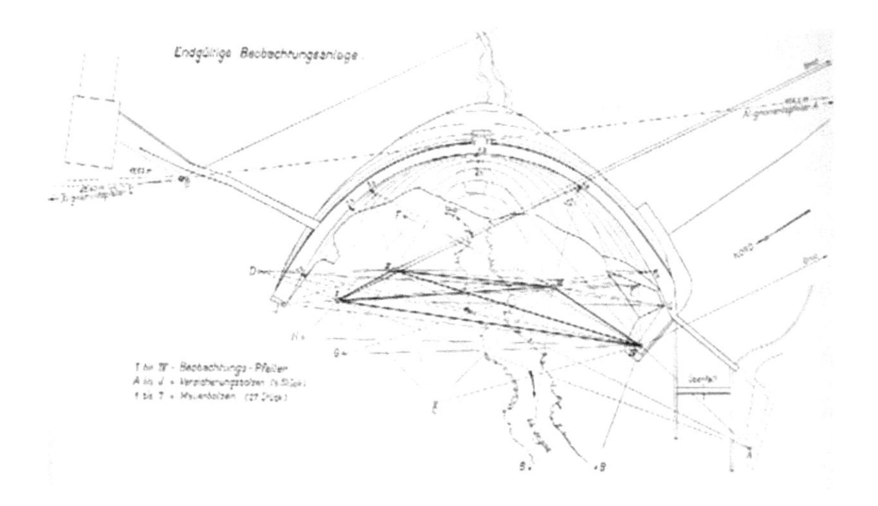


Figure 1. Montsalvens dam Map of the network of trigonometric and alignment measurements (after CSGB 1946).

to measure both the loads which apply on the structure (causes) and the various parameters which characterize the behavior of a restraint (consequences). These measures should already be undertaken during the construction of the structure. Mention is also made of the importance of performing measurements regularly and interpreting them without delay. It is also requested to compile a dam file and keep it up to date. Over time, this regulation has undergone various modifications and improvements. Currently, the legislative aspect is regulated by the federal law on accumulation works (WRFA, 2010. RS 721.102 of October 1, 2010) and its ordinance (WRFO. 2012. RS721.101.1 of October 17, 2012). These were supplemented in 2002 by Regulations which govern the methods of application of the ordinance and in particular provide information on the safety concept. It is based on three pillars which are: structural safety, monitoring and maintenance and emergency plan (Figure 2). It should be noted that the first pillar minimizes the risk while the other two minimize the residual risk.

Monitoring involves setting up a strict organization to monitor the behavior of storage structures and their foundations. It calls on the operator's staff, engineers and experts in the field of dams. They must ensure in all circumstances that the structure and its foundations behave appropriately and, if necessary, propose any useful measure to remedy any potential threat. It should be noted that during the life of the dam, even unpredictable phenomena may occur. Three essential tasks are carried out according to a precise program: a) visual checks, b) direct measurements from monitoring instrument, and c) operating checks of moving parts, instrumentation and means communications.

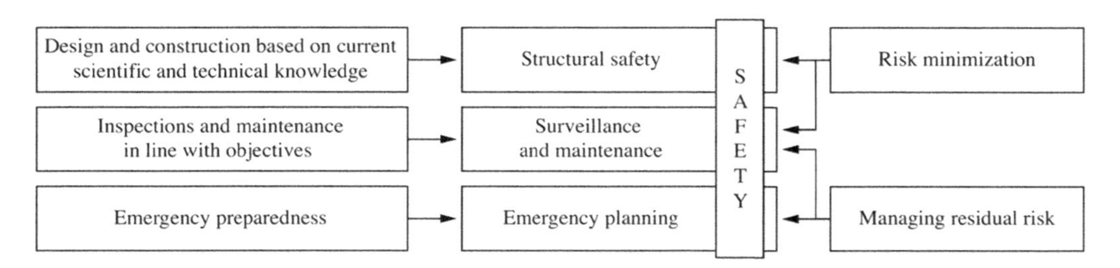


Figure 2. Diagram of the safety concept in effect for dam structures in Switzerland based on the three pillars (from SFOE, 2015a).

## 3 OBJECTIVES OF THE MONITORING SYSTEM

The measurement system has several purposes. First, to run checks during construction and first impounding; for this reason, such system will already be planned as part of the project. Checks are carried out during operation in order to detect any anomalies in behavior in time. Secondly, it permits gaining knowledge on the behavior of the dam and improve the design of new ones.

## 4 PRINCIPLE AND CONCEPT

The monitoring system must permit to measure the external loads which apply to the structure, especially the hydrostatic load (water and ice) and sediment pressure, the temperature of the water and the air and any seismic loads. The various parameters which characterize the behavior of a retaining structure, namely the deformations undergone by the foundations and the body of the dam, the leaks and percolations through the dam and the foundations, the thermal state, the uplift pressure in the dam foundations as well as the pore water pressures and possibly the saturation lines in the dykes (Table 3).

| Concrete dam | Embankment dam | Foundations |
|---|---|---|
| Structural deformation | Deformation in the dam body | Deformation<br>Abutment movements |
| Local movements (cracks, joints) | Specific movements<br>(connection with a concrete structure) | Specific movement<br>(cracks, diaclases) |
| Dam body temperature | Dam body temperature<br>to detect seepage (possible) | |
| Uplift (at the concrete-foundation interface and in the rock) | Pore pressure in embankment dam body and piezometric level | Pore pressure<br>Deep body uplift pressure<br>Piezometric level<br>Groundwater level |
| Seepage and drainage flow | Seepage and drainage flow | Seepage and drainage flow, resurgence (springs) |
| Chemical analysis of seepage water<br>Turbidity (possible) | Chemical analysis of seepage water<br>Turbidity | Chemical analysis of seepage water<br>Turbidity |

Figure 3. Significant parameters measured using monitoring instrumentation (after Schleiss and Pougatsch, 2011).

The monitoring instrumentation project will be adapted to the particularities and the importance of the accumulation structure. In addition, it will be taken into account that the dam and its foundations constitute a whole, so it must make it possible to clearly distinguish the behavior of the dam from that of its foundations and its surroundings. There is no rule to define the number of measuring devices needed to ensure satisfactory monitoring of behavior; it is preferable to have a limited number of reliable instruments, which also facilitates the interpretation of the measurements. To deal with breakdowns or failures, it is recommended to provide redundant measurements of certain parameters (i.e., the measurement of deformations). It should also be noted that a monitoring instrument is not a fixed system. Indeed, it is good to examine periodically if it still satisfies the requirements and the needs; if necessary, it is supplemented, adapted or modernized. Even if the instruments offered are more and more numerous and are constantly evolving, it should however be noted that the parameters to be measured remain the same. Table 4 shows by way of example the different equipment and types of measurement in use for the monitoring of concrete dams and their surroundings.

| Type of measurements | Equipment |
|---|---|
| Structural deformation | Direct pendulum<br>Inverted pendulum<br>Inclinometer<br>Extensometer<br>Fiber-optic sensor and cable<br>Geodesy<br>Geodetic survey (terrestrial measurements and GPS)<br>Leveling<br>Polygonal<br>Vertical line of sight<br>Simple angular measurements<br>Alignment |
| Local movements (cracks, joints) | Jointmeter<br>Micrometer<br>Fiber-optic sensor and cable<br>Dilatometer<br>Deformeter |
| Dam body temperature | Normal thermometer<br>Electronic thermometer<br>Fiber-optic sensor and cable |
| Uplift at the concrete-foundation interface | Manometer<br>Pressure cell |
| Leaks, seepage, and drainage flow | Weir, venturi<br>Volumetric measurements |
| Chemical analysis of seepage water | |
| Tension of anchors (in the body of the dam, in the foundation) | Load cell (hydraulic or electrical system) |

Figure 4. Equipment and types of measurement of a concrete dam.

## 5 REQUIRED QUALITY OF THE INSTRUMENTATION AND EXPECTED EVOLUTION

The choice of measuring devices depends on the parameters to be observed, the construction method of the structure and the installation possibilities. The choice must be adapted to each specific case. Priority must be given to instruments meeting the following criteria (CSB, 2005, 2006): simple and robust,

- precise and reliable,
- durable,
- easy to read,
- insensitive to environmental conditions; provided that they are not integrated into the body of the structure, they will be accessible and replaceable.

To deal with breakdowns or failures, it is recommended to provide redundant measurements of certain parameters (for example, the measurement of deformations). Regarding the reliability of measuring devices, the failure rate is highly variable and depends on the type of instrument. It should also be noted that the longevity of the instruments is less than the lifespan of the dam. In general, suitable monitoring of deformations requires an extensive (Figure 5) and spatial (Figure 6) measuring device, which makes it possible, using geodesy, to collect information on the altimetric and planimetric displacements of selected points. The control or measurement points are located on the crest and in the galleries, on the facings or on the embankments as well as on the ground (surroundings of the dam). As part of the routine monitoring of a low or medium-height structure, we are often limited to monitoring the movements of points located at the crest level. Sometimes a faulty instrument needs to be replaced. It is therefore recommended to have suita-ble reserve instruments (CSB, 2013a).

Figure 5. Extended reference space (terrestrial deformation measurements + GPS) (after Schleiss and Pougatsch, 2011).

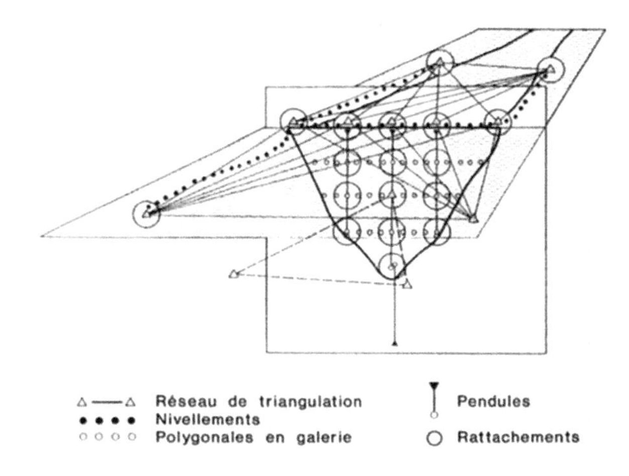


Figure 6. Diagram of a spatial measurement network (after Biedermann, 1997).

## 6 ANALYSIS OF RESULTS AND PREDICTION OF BEHAVIOR

In order to adequately interpret the behavior of a dam, it is possible to rely on mathematical models which determine the expected response under the effect of the applied loads. Interpretive models allow checking the plausibility of the measurements and to identify any irreversible behavior. The two main families of models used are on the one hand statistical and on the other hand deterministic. The latter are particularly useful in the case of the interpretation of the displacements of concrete structures whose reaction is mainly influenced by the level of the reservoir as well as the thermal state.

The overall monitoring process for an accumulation structure is illustrated in Figure 7.

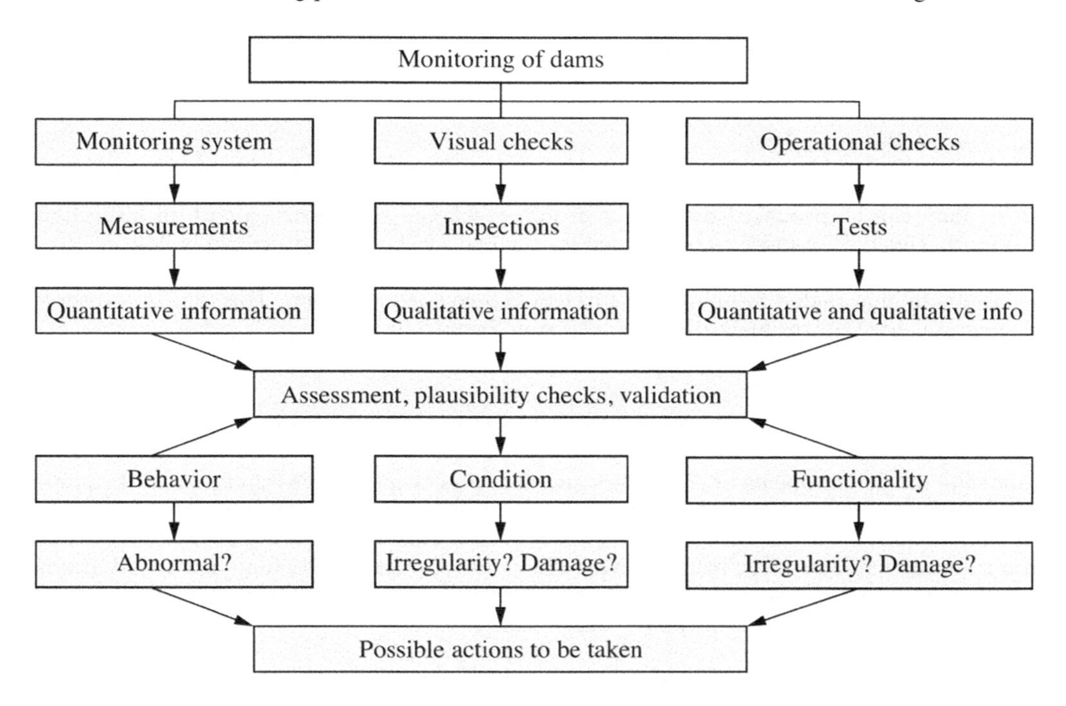


Figure 7. Process for monitoring an accumulation structure (Schleiss and Pougatsch, 2011).

## 7 THE MAIN RECENT DEVELOPMENTS IN INSTRUMENTATION AND THEIR APPLICATIONS

### 7.1 *Strain and temperature measurements*

a) Fiber-optic sensors and cables In the context of new measurement technology, it should be noted that fiber-optics have been used in measuring instruments in dams since the 1990s, primarily as (CSB, 2005a):
   - A device in which the optical fiber is itself the measuring instrument (such as the extensometer)
   - A device in which various phenomena are measured along the fiber-optic cable.
   - A device in which the optical fiber provides a means for transporting data (pressure, temperature, difference in length)

b) 3D measurements of deformation in a borehole.

The borehole micrometer is a mobile measuring device that enables measurement in successive 1-meter sections of, for example, differential variations in length along the borehole. This instrument is equipped with an inclinometer that determines displacement in 3 orthogonal directions along a vertical borehole (CSB, 2005a).

### 7.2 *Displacement measurements*

GPS (Global Positioning System)
Space measurements by satellite (accurate distance measurements between orbits and sensor)

### 7.3 *Various surveys*

a) Ground Survey Aperture Radar (GBInSAR)
Photogrammetry method using ground station images
b) Ground Penetrating Radar (GPR)
Detect changes in properties of near-surface ground layers, localization of defects or voids in concrete structures

### 7.4 *Observation of surfaces*

a) Face observation using a drone A hi-res photographic recording of the downstream face of the dam can be produced by using a drone, which can then form part of an inspection report. This type of inspection was used for the Zeuzier dam in Switzerland in 2016.
b) Laser scanning Laser scanning and digital imagery Accurate distance measurements using laser with high spatial resolution on surfaces (3D geometry of dam). This type of measurement was used for the St-Barthélémy dam A in Switzerland.

## 8 AUTOMATION AND TRANSMISSION OF MEASUREMENTS

Following the developments of electronics and data processing, the possibilities and the interest of the automation of the monitoring instrumentation increased. They allow a direct link with the user. Such devices consist of means of measurement (measuring devices), means of data transmission, automatic means of acquisition and storage of data (databases) and means of processing and presentation of data (analysis of measurement results, drawing up graphics and writing reports - Stucchi, Crapp R., Fern I., 2022).

## REFERENCES

Publications of the Swiss Committee on Dams - SDC (formerly Comité Suisse des Grands Barrages – CSGB or Comité National Suisse des Grands Barrages - CNSGB).

CSGB, 1946. Mesures Observation et Essais sur les Grands Barrages Suisses 1919–1945.

CNSGB, 1964. Comportement des Grands Barrages suisses.

CNSGB, 1985. Barrages suisses - Surveillance et entretien. *published during the ICOLD 15th International Congress, Lausanne, Switzerland.*

CNSGB, 1993. L'informatique dans la surveillance des barrages, saisie et traitement des me-sures. *Groupe de travail pour l'observation des barrages.*

CNSGB, 1997a. Surveillance de l'état des barrages et check-lists pour les contrôles visuels. *Groupe de travail pour l'observation des barrages.*

CNSGB, 1997b. Mesures de déformations géodésiques et photogrammétriques pour la surveillance des barrages, 82 pages. *Groupe de travail pour l'observation des barrages.*

CSB, 2003. Méthodes d'analyse pour la prédiction et le contrôle du comportement des barrages. *Wasser, Energie, Luft - Eau, Énergie, Air, 95e année. cahiers 3/4, 74-98.*

CSB. 2005. Dispositif d'auscultation des barrages. Concept, fiabilité et redondance. *Groupe de travail pour l'observation des barrages.*

CSB. 2006. Dispositif d'auscultation des barrages. Projet, fiabilité et redondance. *Groupe de travail pour l'observation des barrages. Wasser, Energie, Luft - Eau, Énergie, Air, 98e année, cahier 2, 144-180.*

CSB. 2013a. Géodésie pour la surveillance des ouvrages d'accumulation. *Groupe de travail pour l'observation des barrages.*

CSB, 2013b. Instruments de mesures - contrôles et calibrages, *Groupe de travail pour l'observation des barrages* Legislative acts and Regulations.
WRFA, 2010. Water Retaining Federal Act (in French, German, Italian). RS 721.101 of 1st October 2010 (State in 1st January 2013).
WRFO, 2012. Water retaining Federal Ordinance (in French, German, Italian). RS 721.101.1 of October 17, 2012 (state in 1st April 2018).
SFOE, 2015a. Directives sur la sécurité des barrages. PartieA: Généralités. Richtlinie über die Sicherheit der Stauanlagen. Teil A: Allgemeines. In French, German. Articles in Swiss technical journals and Congresses.
Biedermann R., 1987. Anforderung an die Messeinrichtungen von Talsperren. *Wasser, Ener-gie, Luft - Eau, Energie, Air, 79. Jahrgang, Heft 1/2, 10-11.*
Biedermann R., 1997. Concept de sécurité pour les ouvrages d'accumulation: évolution du concept suisse depuis 1980. *Wasser, Energie, Luft - Eau, Énergie, Air, 89e année, cahier 3/4, 63-72.*
Bischof R. et al., 2000. 205 dams in Switzerland for the welfare of the population. *20th International ICOLD Congress, Q.77 - R.64. Beijing 2000, 997-1018.*
Darbre, G. R ., Pougatsch, H. 1993. L'équipement de barrages dans le cadre du réseau national d'accélérographe. *Wasser, Energie, Luft - Eau, Énergie, Air, 85e année, cahiers 11/12, 368-373.*
Deinum Ph. J. 1987. Versuchsinstallation des Sperry-Tilt Sensing Systems zum Erfassen der Durchbiegung der Bogenmauer Emosson. *Wasser, Energie, Luft - Eau, Énergie, Air, 79. Jahrgang, Heft 1/2, 17-19.*
Egger K., 1982. Geodätische Deformationsmessungen. Eine zeitgemässe Vorstellung. *Wasser, Energie, Luft - Eau,Énergie, Air, 74. Jahrgang, Heft 1/2, 1-4.*
Gicot H., 1976. Une méthode d'analyse des déformations des barrages. *12th International ICOLD Congress, Mexico, Vol. IV, communication C1, 787-790.*
Lombardi G., 1992. L'informatique dans l'auscultation des barrages. *Wasser, Energie, Luft - Eau, Énergie, Air, 84e année, cahiers 1/2, 2-8.*
Lombardi G., 2001. Sécurité des barrages - Auscultation. Interprétation des mesures. *Commentaires généraux, 25 pages.*
Pougatsch H., 2002. Surveillance des ouvrages d'accumulation. Conception générale du dispositif d'auscultation. *Wasser, Energie, Luft - Eau Énergie, Air, 94e année, cahiers 9/10, 267-271.*
Sinniger R., 1987. Observation des versants d'une retenue. *Wasser, Energie, Luft - Eau, Énergie, Air, 79. Jahrgang, Heft 9. 209-210.*
Sinniger R., 1985. L'histoire des barrages. *EPFL, Polyrama, 1985, 2-5.*
Schleiss A. et Pougatsch H., 2011. Les barrages. Du projet à la mise en service. *EPFL Press. Traité de génie civil, Vol 17. Presse polytechnique et universitaire romande.*
Stucchi R., Crapp R., Fern I., Development of a Software as a Service platform for dam monitoring, *Hydro 2022, Strasbourg, France.*

# Contributions of geodesy to the safety of dams in Switzerland

A. Wiget
*Former Head of Geodesy, Federal Office of Topography, Swisstopo, Wabern, Switzerland*

B. Sievers
*Prof. em., University of Applied Sciences and Arts Northwestern, Switzerland*

F. Walser
*Senior Engineer and Member of the Board at Schneider Ingenieure AG, Chur, Switzerland*

ABSTRACT: Geodetic deformation measurements have been used successfully in Switzerland for over 100 years to survey and monitor dams. They make it possible to determine any displacements and deformations of the dams as well as the surrounding terrain with high precision and reliability in relation to an absolute reference frame. Combined with other instruments for deformation measurements, geodetic methods contribute considerably to the determination of dam behaviour and to the assessment of exceptional situations or behavioural anomalies of dams and thus to their safety.

This article specifies the tasks and requirements of dam surveys. Established methods of geo-desy, their evaluation as well as recent instrumental developments are described and possible methodological and technological developments are pointed out.

The article published in German in the journal «Wasser Energie Luft» 3-2023 as well as other publications by the working group on dam surveying of the Society for the History of Geodesy in Switzerland (GGGS) can be accessed on the website of the GGGS (www.gggs.ch > Virtuelles Museum > E-Expo Schweizer Talsperrenvermessung). Also available are an extensive bibliography of relevant technical publications beginning in 1920 and a picture gallery from different time-epochs categorized in 17 topics.

RÉSUMÉ: Les mesures géodésiques de déformation sont utilisées avec succès en Suisse pour la surveillance des barrages depuis plus de 100 ans. Elles permettent de déterminer avec une grande précision et fiabilité, par rapport à un cadre de référence absolu, les éventuels déplacements et les déformations des barrages ainsi que du terrain environnant. Associées à d'autres dispositifs de mesure de déformations les méthodes géodésiques contribuent de manière décisive à la détermination du comportement des barrages et à l'évaluation des situations exceptionnelles ou des anomalies de comportement des ouvrages d'accumulation, et donc à leur sécurité.

Cet article explique les tâches et les exigences des levés de barrages. Les méthodes éprouvées de la géodésie, leur évaluation ainsi que les développements instrumentaux récents sont décrits et les développements méthodologiques et technologiques possibles sont soulignés.

L'article publié en allemand dans la revue « Eau énergie air » 3-2023 ainsi que d'autres publications du groupe de travail sur la mensuration des barrages de la Société pour l'histoire de la géodésie en Suisse (SHGS) sont disponibles sur le site web de la SHGS (www.gggs.ch > Virtuelles Museum > E-Expo Schweizer Talsperrenvermessung). On y trouve également une vaste bibliographie de publications spécialisées depuis 1920 et une galerie d'images sur 17 thèmes couvrant toutes les époques.

## 1 TASKS AND REQUIREMENTS OF DAM SURVEYS

According to the "Directive on the Safety of Water Retaining Facilities" issued by the Swiss Federal Office of Energy, geodetic measurements are an integral part of dam surveillance. Used in combination with other means and instruments for detecting deformation, they contribute to:

- the determination of the behaviour of dams (as part of the ongoing assessment of the impacts and conditions of the structure);
- rapid assessment in case of extraordinary situations or following an extraordinary occurrence;
- clarification of causes of anomalous behaviour detected by other measurement instruments.

Geodetic measurements can be used as a stand-alone method to determine the deformation and displacement behaviour of dams and reservoirs and their surroundings. However, they are usually used in conjunction with other measurement systems, such as pendulum systems, to determine changes in the position and height of selected points on a dam, thus providing redundancy in the overall monitoring concept. The control points included may be located on the crest and at different elevations at the airside surface (downstream face) of the dam; and, if accessible through galleries, within the structure (e.g. reference points of pendulum measurements or of traverses in the galleries), on the banks and in rock formations in the immediate vicinity, as well as in the wider environment of the dam outside its pressure zone. Finally, critical terrain features such as landslide slopes or glaciers in the danger zone of the dam can also be monitored. The magnitude of the accuracy requirements can be summarised as follows:

Table 1. Accuracy requirements for dam surveys.

| Objects of the survey<br>Deformations | arch or gravity dam<br>(VA or PG)<br>(concrete dam)<br>mm | earth or rock fill dam<br>(TE or ER)<br>(embankment dam)<br>mm | environment, critical<br>terrain zones<br>(e.g. landslides)<br>mm |
|---|---|---|---|
| Horizontal | 0.5 – 1 | 2 – 5 | 5 – 10 |
| Vertical | 0.1 – 0.2 | 0.5 – 5 | 5 – 10 |

The Subgroup Geodesy of the Working Group on Dam observations of the Swiss Committee on Dams has developed detailed recommendations for the use of geodetic deformation measurements at dams (Schweizerisches Talsperrenkomitee STK/CSB, Arbeitsgruppe Talsperrenbeobachtung 2013). Long-term reliable deformation measurements with millimetre accuracy are a complex and challenging field of application for engineering geodesy; and the observations are time consuming. They must be carried out and evaluated by specialists with the necessary competence and experience, using high quality and tested instruments (Walser 2014). Furthermore, in addition to civil engineering knowledge of the possible dam behaviour, an understanding of geological and geotechnical aspects is required. Geodesy provides the basis for determining the deformations and displacements of the dam in combination with the control procedures and measurement methods of the dam operators, such as the visual inspections, clinometer, extensometer and pendulum measurements, which monitor the geometric behaviour of the structure itself. The results are used by the civil engineering and geology experts to judge the slide safety of the dam and to verify its stability. As part of the dam safety inspection, geodetic measurements are usually carried out at least every five years. The reports of the geodetic deformation measurements are therefore kept as part of the file collection on the dam facility.

## 2 PROVEN METHODS AND INSTRUMENTS OF GEODESY

Geodetic deformation measurements on dams have a history of more than 100 years in Switzerland (Wiget et al. 2021, Wiget 2022). The oldest method of monitoring dams is geometric alignment. Starting from a stable pillar assumed to be fixed, a vertical plane is established by aiming the alignment instrument at a reference mark (target). Using the instrument's telescope, the horizontal deviations of the alignment points on the crest of the dam (signalled by alignment target marks or by means of a measuring rod) are measured from this vertical plane.

The accuracy and reliability of alignment observations were limited by refraction phenomena and uncontrolled stations or pillars (assumed to be fixed). On the occasion of the construction of the *Montsalvens* dam (canton FR) in 1921, which was the first double arch (horizontally and vertically curved) dam in Europe, engineers of the Swiss Federal Office of Topography (swisstopo) proposed the application of trigonometric methods used in national surveying: direction and angle measurements (triangulation) and precision levelling. For this purpose, theodolites were used for repeated bearing intersections (forward intersections, see Figure 1) from (at least) two observation pillars outside the dam to target control points on the crest and on the downstream airside surface of the dam, in combination with more distant reference points ("fixed points"). To determine the grid scale, the distance between the observation pillars had to be measured. In addition to the horizontal position observations, the target points could also be monitored vertically by means of height angle measurements. The first measurements were made at Montsalvens in January 1921, before the initial filling of the reservoir, and in November 1921, when the lake was full.

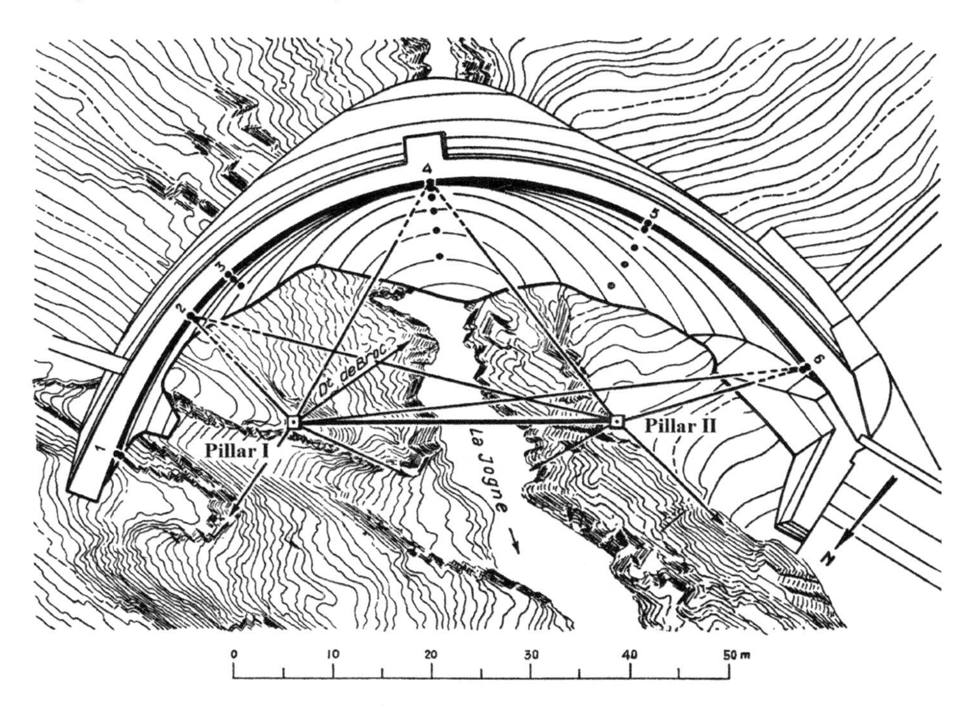


Figure 1. Forward intersections using angle measurements at Monsalvens dam, 1921.

Geodetic deformation measurements were also used at the *Pfaffensprung* dam (canton UR) as early as 1922. In order to be able to measure dam deformations and movements in real time during filling or emptying of the reservoir, all points were measured simultaneously by two observers from two pillars by means of forward intersections.

During the first epoch of Swiss dam construction until the middle of the 20th century, swisstopo was the only institution in Switzerland to carry out geodetic deformation measurements on dams. With the upswing in Swiss dam construction from 1950 onwards, the geodetic methods of dam monitoring were further developed and taught at the Swiss Federal Institutes of Technology in Zurich and Lausanne (ETHZ and EPFL). As a result, engineers from private surveying offices were increasingly commissioned to carry out this work, which eventually also became essential due to the large number of objects to be monitored.

Geodesy has been used for surveillance of all major dams in Switzerland from the very beginning. The methods, of course, have been continuously improved and adapted to technological developments. The measurement networks have been extended and the number of well-founded observation pillars has been increased, partly because of the increasing size of the dams, but also in order to be able to better monitor the stability of the pillars and reference points. Therefore, some of the observation pillars are located up-stream and downstream of the dam, outside the zone of load influence (pressure bulb). Measurements are taken from the 'fixed' to the 'moving' and give 'absolute' displacements in relation to the chosen reference frames. Whereas in the early days the focus was on short-term differential movements between two epochs, today's measurements are designed for long-term studies.

In addition to the trigonometric measurements already mentioned, precision levellings are measured in order to monitor the changes in height (subsidence, rise) of the dams or of individual parts of the structure, as well as the foundation and the banks in their vicinity, which are also of great interest (Figure 2). The levellings provide even higher accuracies than trigonometric height differences. They usually run from reference points on one side of the valley over the crest to the other side, but are also measured in the galleries within the dam. Wherever possible, the height reference points are located in stable areas that are not subject to the pressure effects and load influences of the reservoir's varying water volumes, and if necessary in geologically stable areas further away.

Figure 2. Precise levelling in the vicinity of Contra dam (canton TI).

Today, the "classical" methods (trigonometry and levelling) still form the backbone of geodetic deformation measurements on dams. However, the geodetic methods have undergone continuous development in terms of measuring stations (e.g. pillars and target bolts) and instrumentation (theodolites, levelling instruments, forced centring, etc.). Since the 1970s, developments in electronics and instrumentation have contributed to a significant increase in the accuracy of geodetic deformation measurements. The biggest step forward was the development of electro-optical distance measuring devices. Previously, distances were measured between the survey pillars by precision invar subtense bars. In polygonal traverses over the dam crest and in galleries invar wires are still used today. Since 1973, electro-optical distance-measuring instruments of the highest accuracy class have been on the market (e.g. Kern Mekometer ME3000 and ME5000). These allowed distances to be measured in the sub-millimetre range in the immediate vicinity or inside the dam, and in the millimetre range over several kilometres in the outer extended network, provided that the representative meteorological parameters are carefully recorded. These instruments were convincing not only because of their high measurement accuracy, but also because of the relatively short measuring time of a few minutes. Today, modern electronic total stations, i.e. theodolites with built-in distance meters, provide similar accuracies with even shorter survey times.

The introduction of electronics brought further important innovations to geodetic instruments: Automatic levelling of the instruments, considering the inclination of the vertical axis; electronic readings of the circle of the theodolites or total stations; automated reading of the staff for digital levels; support of the data acquisition and preprocessing on site by means of external software; digital recording of measurement data on internal or external data carriers, etc. Thanks to developments such as motorisation and automatic reflector alignment, the latest generation of instruments allow faster and more convenient measurements, which in turn can have a positive effect on accuracy. They also enable automated measurement systems for continuous monitoring of dams and their surroundings, known as geodetic monitoring systems (see Geomonitoring in chapter 4).

Since the late 1980s, terrestrial measurements have been supplemented by GPS measurements, today known as GNSS measurements, incorporating all available Global Navigation Satellite Systems. As these do not require line-of-sight between the stations or points to be measured, the external measurement networks can be extended to more distant reference points in geologically stable zones, unaffected by the water retaining facility and its load influences. Thus, long-term displacements of the dam and the surrounding area can be monitored in the well controlled three-dimensional reference frames in the range of a few millimetres.

Especially for large dams, geodetic measurements are used in combination with other measurement systems such as pendulum measuring systems, extensometers, joint meters or deformation meters, etc. These usually provide relative displacements and deformations, whereas geodesy measures absolute displacements in position and height with respect to the above-mentioned reference frames. For the interpretation of the behaviour of the dam and its surroundings, an optimal linking of the so-called "inner" and "outer" measuring systems is therefore essential. The analysis must take into account the local position as well as the temporal execution of the measurements (time, frequency), with careful registration of the respective environmental conditions such as water level or air and concrete temperatures. For example, pendulum system reference points (set plates and suspension or anchor points) should be connected directly or indirectly to the geodetic measurement network. Finally, geophysical and geotechnical instruments such as sliding micrometers or deformation meters, borehole extensometers, etc. can also be linked carefully and precisely to the geodetic network.

Table 2. Advantages and qualities of geodetic deformation measurements.

- Geodetic measurement methods are very adaptable, from the installation of the network, the instrumentation, signalisation and execution of the measurements, to the evaluation; they can be well adapted to the different types of dams and local conditions.
- Geodesy makes it possible to determine the absolute deformations and displacements of dams and their surroundings in relation to the immediate and wider environment and, if the measurement networks are extended accordingly, also to regional or national reference frames in geologically stable or well-studied areas.
- Movements of dams can be recorded even if they are not associated with changes in slope, strain or stress.
- Measurements are made in one, two or three dimensions (horizontal and vertical) and deformations can be analysed accordingly.
- Short-term (elastic) changes, e.g. during the initial accumulation of the reservoir, can be determined as well as long-term phenomena (e.g. subsidence, rise due to concrete changes) or permanent deformations and trends in the measurement series over decades, even with changed methodology and renewed instrumentation.

## 3 EVALUATION AND DEFORMATION ANALYSIS

The goal of geodetic deformation measurements is to monitor and record the kinematic behaviour of dams, its foundation and surrounding by taking "snapshots" of its geometry at different time epochs, calculating changes of control points in position and height together with their corresponding accuracy, and describing the differences or movements in an appropriate way. The kinematics to be studied may vary. For example, relatively short-term movements or reversible deformations of the dam in the context of filling or emptying of the reservoir may be of interest, or the long-term stability of the dam over decades may be to be investigated. The questions to be answered also influence the conditions under which the measurements must be carried out and how they are evaluated.

The first trigonometric measurements were evaluated "by hand" using relatively simple functional models or graphical methods. Today, digital recording or on-line data transfer to the analysis software in the case of automatic measuring systems and electronic data processing allow faster and less error-prone evaluation of geodetic measurements and the adjustment of extensive measurement networks. This and the subsequent deformation analysis can be roughly divided into the following phases:

1) Verification of the measurements carried out by specific processing of the raw measurements according to the measurement procedures (directions, angles, distances, levelling, GNSS), taking into account calibration values, meteorological conditions as well as geometric corrections such as instrument heights, etc.; modelling and correction of systematic error influences; determination of stochastic key figures for parametric estimation.
2) Network adjustment of all measurements of an epoch to calculate the epoch-specific geometry (coordinates, heights) of the permanently marked control points on the structure and in the terrain; "fixed point analysis": i.e. testing of fixed point hypotheses and reasonable selection of fixed points in the chosen reference frame. The calculated coordinates and heights of the control points, as well as their accuracy, refer to the selected fixed points.
3) Overall adjustment of all measurements and point calculations over several epochs or adjustment of all previous measurement epochs, considering the above mentioned fixed point analysis (the aim is to achieve a spatial reference frame as uniform as possible in the long term); calculation of the epoch-specific coordinates and heights in all epochs in relation to the selected reference points, considering their accuracies.
   In addition to point coordinates and heights, the adjustment provides a great deal of additional information which is important for assessing the quality of the measurements and for analysing displacements and deformations (see also Table 3):
   - Accuracy in the form of empirical standard deviations and confidence ellipses: The values are "absolute" with respect to the fixed points, but "relative" between the determined points in the same epoch or in different measurement epochs;
   - Information on the geometric reliability of the coordinates and heights;
   - Estimation of the achieved overall accuracy of the different types of observations (so-called variance component estimation).
4) Deformation analysis: Calculation of the position and height differences of the control points over one or more periods (difference between two epochs), including their accuracy and reliability; i.e. calculation of short-term differences, displacements or deformations, e.g. between the last two measurement epochs, as well as comparison and determination of long-term trends over all epochs; reporting of relative accuracies compared to the selected reference points as well as between the measurement epochs; clarification of the significance of the changes for the selected confidence level (e.g. 95% or 99%) for the detection of real displacements.
5) The measurements carried out (including measurement programme, personnel, instruments and measurement conditions), the evaluation procedure, the fixed point analysis and the results of the deformation analysis are documented in a technical report as part of the long-term monitoring of the dams. The results are presented in clear tables and descriptive graphs and are evaluated from a geodetic point of view (including choice of control points, significance, special conditions before and during the measurements) as a basis for the technical assessment by the experts (Figure 3).

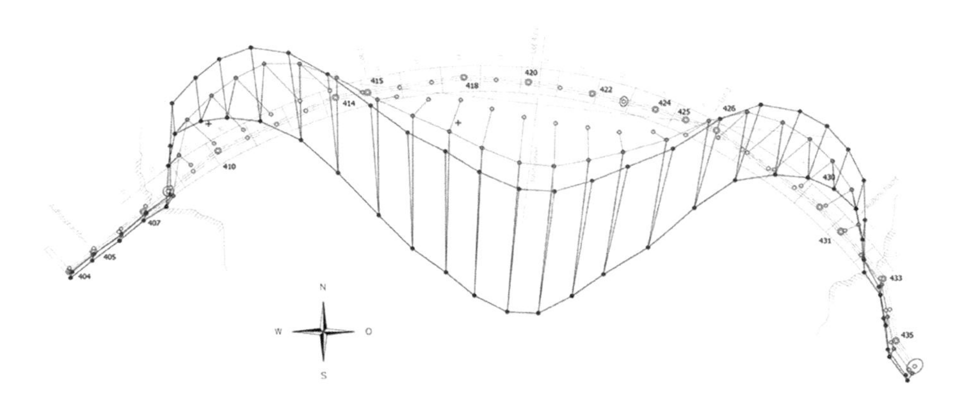


Figure 3. Horizontal displacements over four epochs.

Table 3. Quality features of the evaluation of geodetic deformation measurements and results.

- Adjustment of the precisely collected and recorded measurements using high quality functional (physical) models and considering verified stochastic parameters of the measurement methods used.
- Determination of the coordinates and heights of the control points on the dam, inside the structure and in the surrounding area to be investigated, with high precision (relative accuracy) as well as absolute accuracy with respect to the reference points, which can be located at almost any distance from the structure, defining the reference frame.
- Optimal combination with other measurement methods to determine displacements and deformations of the dam and surrounding terrain with respect to the reference frame.
- Controlled high reliability of all results.

## 4 MODERN METHODS AND OUTLOOK ON POSSIBLE FUTURE TECHNOLOGICAL DEVELOPMENTS

The methods described so far enable surveying of discrete individual points (control points) at periodic intervals and comparing them with previous surveys. Differential, irregular movements in space and time between the surveyed points are usually not determined; they can be interpolated if necessary.

But with today's instruments and software tools, automated, continuous monitoring of the position and height of dams and their surroundings is possible. For permanent monitoring with high accuracy requirements and temporally dense sampling rates, geodetic monitoring systems (so-called geomonitoring) are increasingly being used. A well-known example was the monitoring of the dams above the Gotthard Base Tunnel during its construction. Such geomonitoring applications are conceivable, for example, for monitoring critical terrain zones like slopes and unstable parts of the valley that could damage the dam structure or its auxiliary works directly or indirectly by falling into the reservoir. Dams of pumped storage power plants, which are additionally or increasingly stressed by more frequent and rapid changes in lake levels, are particularly suitable to be automatically and permanently monitored by monitoring systems. Temporary monitoring systems are installed especially when work or construction has to be carried out on, under or in the vicinity of the dam that can endanger its safety.

The monitoring systems can be composed of several different sensors, such as total stations, digital levels, GNSS receivers, electronic inclinometers and hydrostatic levelling, CCD sensors, inertial measurement units (IMU), extensometers, temperature and pressure sensors or other modern measurement methods, which will be mentioned below (TLS, In-SAR, FOS, etc.). The software systems for control, evaluation, analysis and alarms are integrated into the monitoring system. The results are determined in real time, broadcasted or made available for viewing via web browser; alarms and error messages are transmitted at critical moments.

New techniques for area and time continuous measurements of deformations are laser scanning and radar interferometry. They shall be briefly described:

Terrestrial laser scanning (TLS): The scanner measures the dam surface in a freely definable geometric point grid (in horizontal and vertical direction) in three dimensions without contact, providing oblique distances, intensity and possibly RGB colour values (Barras 2014). The rapid measurement of large numbers of points (up to 1 million points/second) is less precise (a few mm to cm) than the conventional, highly redundant but time-consuming multi-point determination (sub-millimetre to mm). However, the mentioned accuracies of the displacements can practically only be determined one-dimensionally in the direction of the laser beam. In addition, the georeferencing and modelling of the point cloud is a challenge. In the future, but probably for some time to come, TLS will increasingly be used in combination with traditional geodetic methods.

Terrestrial and satellite radar interferometry: In the case of high risk potential, terrestrial radar interferometry (ground-based interferometric synthetic aperture radar, GB-InSAR) can monitor surface deformations at dams or in their vicinity, quickly over a wide area or, if necessary, continuously with millimetre precision (Jacquemart & Meier 2014). The range of the sensor is up to 4 km, the area covered is over 5 km$^2$. Movements can be detected in the millimetre range, under favourable conditions even in the sub-millimetre range, but only in the direction of the axis of the radar beam (line-of-sight LOS). By using several GB-InSAR sensors at different stations, 3D displacements can be determined. So far, the method has only been used experimentally in the vicinity of Swiss dams.

Interferometric Synthetic Aperture Radar (InSAR) from satellites can be used to monitor the surface of entire valleys or countries. The areas that can be measured and the periodicity result from the illumination zones of the satellite passes. In Switzerland, for example, subsidence in salt mining areas, landslides, block glaciers and permafrost areas are studied. The accuracy of the average displacement rates is better than 1 mm/year in the LOS direction and better than 4 mm/year for individual measurements.

Table 4. Other methods related to or combined with geodesy.

- Close-range photogrammetry: High quality photographs, if necessary with overlaps for stereographic evaluation, to document and interpret cracks and other surface changes. Today, drones (unmanned air vehicles UAV) are also used as sensor carriers (Figure 4). Accuracies are in the centimetre range. This area of application is currently developing rapidly and it can be assumed that the use of artificial intelligence will lead to major increases in productivity.
- Deformation Camera: Automatically analyses sequential high-resolution imagery and uses sophisticated image processing techniques to determine two-dimensional deformations of unstable slopes, rock faces or rock glaciers to an accuracy of a few centimetres.
- Monitoring of deformations in dams using integrated fibre optics with integrated sensor systems (FOS) to determine changes in length, e.g. in block joints, with an accuracy of a few micrometres.
- Rockfall radar: Detects rockfall events in all weather conditions, including darkness, and alerts within seconds.
- Digital geotechnical sensors for sub-millimetre fracture measurements (extensometers, telejointmeters, etc.).
- Motion sensors, piezoelectric sensors or MEMS (Micro-Electro-Mechanical Systems).
- Digital level measurements.

High-precision, long-term reliable deformation measurements are an exciting, complex and highly demanding field of application for engineering geodesy. Geodetic monitoring of dams will remain an important pillar in the safety concept of dams, mainly because of its "absolute" results.

Figure 4. UAV for close range photogrammetry of dam surface.

Table 5. Possible future developments.

- Measurement systems with increased networking and integration of geodetic, geotechnical and other, possibly new, sensors (meteorological, inertial, tide gauges, etc.).
- A transition from periodic measurements to continuous time series at selected, permanently installed monitoring stations, thanks to lower sensor prices even in large numbers.
- Integration of the geodetic dam monitoring networks by means of GNSS into the "absolute", well-monitored and long-term stable reference frame of the national survey for inter-regional comparisons, e.g. in case of earthquakes.
- Evaluation and analysis tools with advanced algorithms, i.e. more complex adjustment methods, near real-time 3D time series and strain analysis, trend derivation, cloud services, artificial intelligence, deep learning.
- Use of new Internet of Things technologies for networking and remote control of autonomous multi-sensor systems (machine-to-machine communication via 5G, IPv6).
- Terrestrial positioning systems using pseudolites (pseudo-satellites, i.e. locally mounted microwave transmitters), analogous to Ground Based Augmentation Systems (GBAS) in aviation.
- Technologies from indoor navigation methods.
- Modern representation methods and graphic tools such as augmented and virtual reality for simulating deformation processes or predicting future object states.

## REFERENCES


Barras, V. 2014. Lasergrammétrie terrestre – une Solution pour l'Auscultation Surfacique. *Wasser Energie Luft* 106(2): 112–115.

Jacquemart, M. & Meier, L. 2014. Deformationsmessungen an Talsperren und in deren alpiner Umgebung mittels Radarinterferometrie. *Wasser Energie Luft* 106(2): 105–111.

Schweizerisches Talsperrenkomitee STK/CSB, Arbeitsgruppe Talsperrenbeobachtung 2013. Geodäsie für die Überwachung von Stauanlagen. See also publications and contributions to CSB/STK symposia 2014 and 2022. https://www.swissdams.ch/de/publications/journees-d-etudes.

Walser, F. 2014. Geodäsie für die Talsperrenüberwachung. *Wasser Energie Luft* 106(2): 101–104.

Wiget, A., Sievers, B., Huser, R. & Federer U. 2021. Beiträge der Geodäsie zur Talsperrensicherheit – Zum 100-jährigen Jubiläum der Talsperrenvermessungen in der Schweiz. *Geomatik-Schweiz* 119 (7/8):170–177 in German, 178-185 in French.

Wiget, A. 2022. Hundert Jahre Talsperrenvermessungen in der Schweiz. *Wasser Energie Luft* 114(1): 39–42.

# Swiss contribution to geology and dams

A. Jonneret, T. Bussard & G. Schaeren
*Norbert SA, Switzerland*

ABSTRACT: Geology is a key factor in the site selection, design, construction and safety of dams. This article summarizes the different geological aspects that must be considered at the various stages of a dam's life. The foundations and abutments, integral parts of a dam, are then discussed. Among other things, they must ensure the stability of the structure, control and limit water seepage and pressure build up below the dam, resist internal erosion and not deteriorate during the life of the structure. The geology of the reservoir and of the catchment area is also an essential aspect to consider in the design and surveillance of dams. In particular, the stability of the slopes must be studied in order to assess the associated hazards (propagation of unstable masses and creation of an impulse wave, blockage of emergency organs, etc.). Targeted investigations, necessary during the design of the dam, monitoring of excavations during construction and a detailed surveillance plan for the life of the structure must be put in place in order to identify any abnormal evolution and, if necessary, to define appropriate mitigation measures. It is essential that the geological model is communicated in a clear and pragmatic way to the various project stakeholders.

RÉSUMÉ: La géologie est un facteur clé pour l'implantation, la conception, la construction et la sécurité des barrages. Cet article résume les différents aspects géologiques qui doivent être pris en compte aux différentes étapes de la vie d'un barrage. Les fondations et les appuis, parties intégrantes d'un barrage, doivent notamment assurer la stabilité de l'ouvrage, contrôler et limiter les infiltrations d'eau et les sous-pressions sous le barrage, résister à l'érosion interne et ne pas se dégrader pendant la durée de vie de l'ouvrage. La géologie du réservoir et du bassin versant est également un aspect essentiel à prendre en compte dans la conception et la surveillance des barrages. En particulier, la stabilité des pentes doit être étudiée afin d'évaluer les dangers associés (propagation de masses instables et création d'une vague d'impulsion, blocage des organes d'urgence, etc.). Des investigations ciblées, nécessaires lors de la conception du barrage, un suivi des excavations pendant la construction et un plan de surveillance détaillé pour la durée de vie de l'ouvrage doivent être mis en place dans le but d'identifier toute évolution anormale et de définir, le cas échéant, des mesures de mitigation appropriées. Il est essentiel que le modèle géologique soit transmis de façon claire et pragmatique aux différents acteurs liés à l'ouvrage.

## 1 INTRODUCTION

Dams are complex structures that require careful design, construction, and surveillance to ensure their long-term stability and safety. Statistics (ICOLD, 1995) indicate that most dam failures occur in newly built dams with approximately 70% of dam failures happening within the first ten years. When it comes to concrete dams, foundation problems are the leading cause of failure, with internal erosion and insufficient shear strength contributing equally (each 21%) to failures. On the other hand, embankment dams face different challenges, with overtopping being the most common cause of failure, accounting for 49% of cases. Internal erosion within the dam body follows closely at 28%, while foundation erosion accounts for 17% of failures in this type of dam. These statistics emphasize that the geology is a sensitive issue, and that the foundation is to be considered as an integral component of the dam. Consequently, appropriate assessment and monitoring should be conducted not only for the dam foundation but also for the slope stability of the reservoir and above the dam crest to mitigate risks including the potential generation of impulse waves.

To this effect, the Swiss Directive on Dam Safety provides a comprehensive set of guidelines and regulations aimed at ensuring the safety of dams in Switzerland. It is based on three elements: structural safety, surveillance & maintenance and emergency response plan. It is enforced by continuous surveillance, yearly detailed report and five yearly comprehensive reports carried out by two experienced independent experts including a civil engineer and a geologist (Swiss Federal Office of Energy SFOE, 2015).

This paper will therefore review the main geological aspects which have to be taken into consideration throughout the life span of a dam to ensure its long-term safety.

Establishing and updating a geological model is the thread that runs through the various geological studies and assessments from feasibility to long-term surveillance. To this end, the importance of a sound geological investigation, construction supervision and geotechnical monitoring will be discussed. An overview of the main geological and hydrogeological aspects that may affect a dam foundation and its reservoir is given.

## 2 GEOLOGICAL MODEL FROM FEASIBILITY STAGE TO DAM SURVEILLANCE

As it has been shown in the previous chapter, the geological aspects have an essential impact on dam design and safety. It is therefore important to review the best practice to ensure that all these essential aspects have been considered from the feasibility study stages throughout the entire life span of the dam. A description of the various geological related tasks which need to be carried out at various stages of a project is provided below.

### 2.1 *Feasibility studies*

During the prefeasibility stage of dam projects, a thorough evaluation is conducted considering various selection criteria such as cost, socio-economic and environmental consideration as well as geographic and geological aspects. These include the topography, the geology of the foundations, the tectonic setting, the availability of construction materials, the watertightness of the reservoir, and natural hazards both at the dam site and around the reservoir. In general, several options of dam sites are assessed at this early stage in order to determine the most suitable site.

The geological assessment should begin with a desk study, which involves reviewing existing geological and geotechnical data available for the foreseen site. This includes geological maps, geological reports, previous drilling or exploration data, seismic activity records, and any other relevant geological information. This will be followed by geological mapping of the site and its surrounding area in order to identify rock types, structural features (faults, folds, fractures), and any potential geological hazards. Preliminary geological and hydrogeological investigations are then conducted to assess the depth of bedrock, its weathering, the extent in depth of decompression, the engineering properties of the soils and rocks at the site, the orientation, nature and geotechnical characteristics of discontinuities and the hydrogeological conditions at the future dam site. This typically involves a geophysical campaign, pitting and trenching, drilling cored boreholes, carrying out in situ tests such as Lugeon water tests and collecting samples for laboratory testing. The investigation can also be complemented by a detailed assessment of the topography and rock slopes by photogrammetry and LiDAR measurements.

A site-specific seismic hazard assessment is conducted in collaboration with specialists which involves studying historical seismic data, analysing fault lines, assessing the potential for ground shaking, topographic and site amplification, liquefaction, landslides, or other seismic hazards that could affect the dam's stability including reservoir-triggered seismicity

(RTS). In addition, the behaviour of the foundation in case of an earthquake must be assessed to ensure that failure leading to uncontrolled water flow due to earthquake loads can be excluded.

Finally, an assessment of the geological hazards, such as landslides, rockfalls, or other geological events that could affect the dam's structural safety and operation are assessed.

Based on the findings of the geological investigation, a comprehensive report is prepared. The report includes a summary of the geological and geotechnical conditions, potential geological hazards, design considerations and recommendations for further studies. This report therefore includes the first geological model which will be tested and updated by additional studies at later stages of the project development.

Figure 1 is an example of a geological cross section compiled for the feasibility study of the new Fah dam in Switzerland.

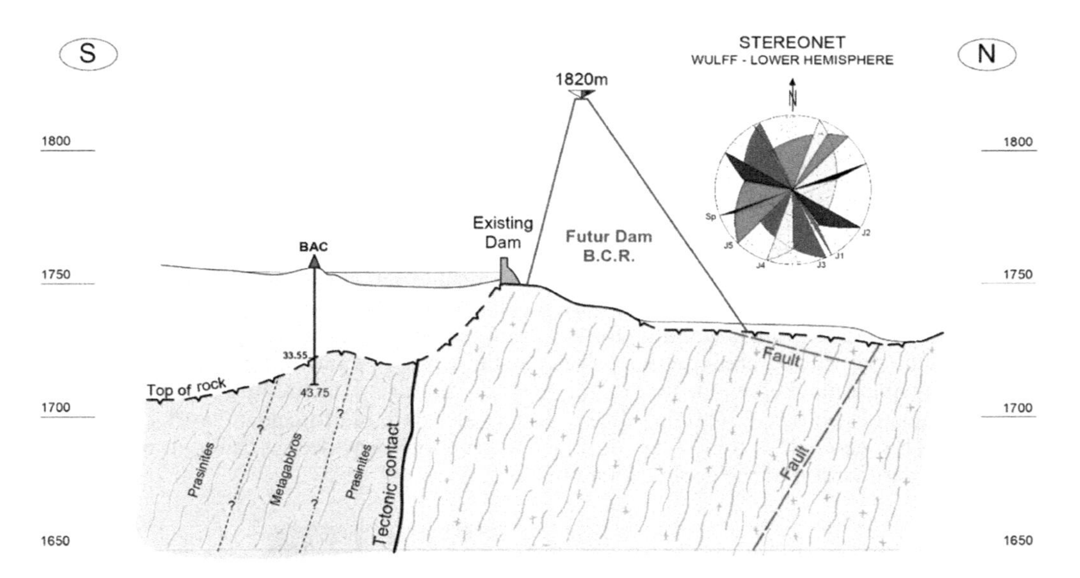


Figure 1. Illustration of an interpretative geological cross section (RCC-BCR dam project in Switzerland).

## 2.2 *Design stage*

Once the site of the dam has been defined and the design is ongoing, it is essential that an update of the geological model is carried out to ensure that all geological aspects which could have an impact on the design of the dam and its appurtenant structure have been reviewed. At this stage, additional detailed geological mapping and complementary investigation whether geophysical or by the means of boreholes and possibly exploratory galleries will be required. Good collaboration and interaction with the engineer responsible for the design are essential. The geological model will then be updated and will include all the assumptions made for the design. At this stage a 3D geological model is a very effective tool that can be used by the engineer for design as illustrated in Figure 2. The successful project should identify geological hazards and, if necessary, mitigate the potential effects that hazards could bring on the construction and operation of the dams and reservoirs. It is therefore essential that the engineers develop a clear understanding of the geological model delivered by geologists (World Bank, 2021).

## 2.3 *Construction*

Geological follow-up of the construction phase is essential to control that the geological model developed during the design phase is adequate. Any unexpected geological conditions

Figure 2. Illustration of a 3D Geological model prepared for the design of a large dam (in yellow) in Central Asia. It includes all the geological mappings (underground and surface), geophysics, boreholes and in situ tests as well as the dam and appended structures.

encountered during excavation should be promptly assessed, and appropriate measures taken to address them whether by additional treatment of a zone or adaptation of the design.

In addition, a detailed mapping of all the dam and appurtenant structures' excavations should be carried out together with records of geological observations and of mitigation measures which might have been taken. Not only will it provide the basis on which the geological model can be updated but it will also form a comprehensive record of the geological conditions encountered during construction. These as-built records are essential in assessing the cause of any abnormal behaviour that might be observed during the life of the dam.

An example of geological mapping during Les Toules dam reinforcement works is given in Figure 3. This geological follow up during construction provided as built drawings and allowed the verification of the geological model.

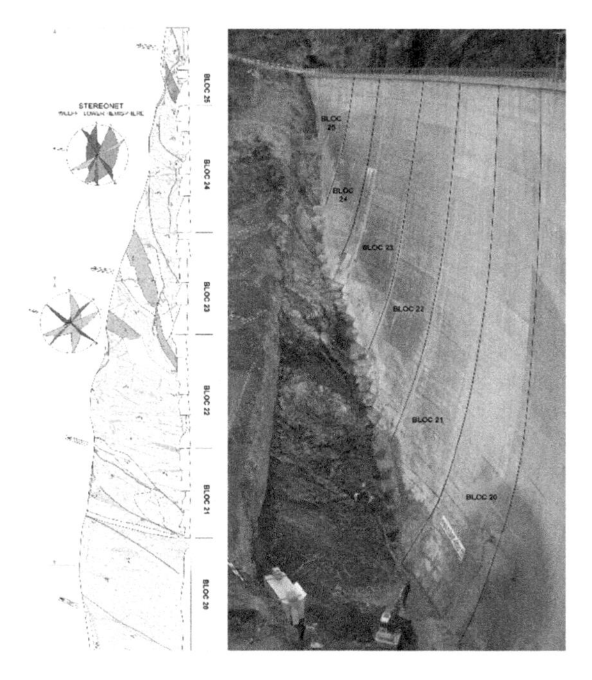


Figure 3. Geological as-built mapping during reinforcement works at Les Toules dam - Switzerland.

2.4 *Surveillance plan*

Visual observation and monitoring the long-term behaviour of a dam is crucial to control its stability, performance, and safety. Geotechnical monitoring of a dam typically involves the use of various instruments to assess the behaviour and stability of the dam and the reservoir area. The most commonly used instruments are inverted pendulums, extensometers, piezometers, gauging stations (discharge of drainages), settlement gauges, strain gauges, inclinometers, seismic monitoring equipment. In addition, chemistry monitoring of possible water seepage can be implemented, in particular when soluble rocks prevail in the dam foundation and abutments (monitoring of the evolution of the water concentration and dissolved mass flow).

Figure 4 illustrates the example of 2 dams, part of the Inga hydro scheme in DRC, Africa where long-term irreversible deformation was observed (Droz et Wohnlich 2019). Rehabilitation of existing instruments was carried out and additional instrumentation put in place to ensure a state-of-the-art monitoring of the dams and power houses.

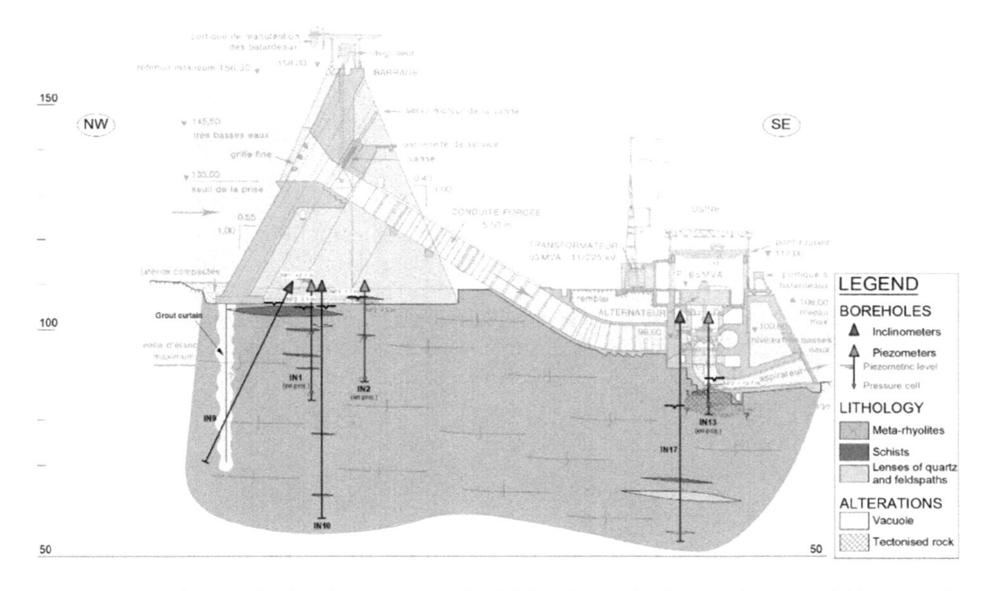


Figure 4. Illustration of the implementation of additional monitoring equipment within one of Inga dam and powerhouse fondations which includes piezometers (in green) and chains of extensometers and inclinometers (in red) – DRC, Africa.

Another example is the Lessoc dam in Switzerland, where monitoring of the dam confirmed the concerns about the stability of the rock spur on the left bank and the weakness of the grout curtain revealed by piezometric measurements. Reinforcement works were carried out in 2019, including the strengthening of the rock spur with 33 passive anchors, the extension of the grout curtain on the left bank and the installation of new monitoring instruments (Fern, 2019).

In addition to all the instruments installed in the dam body or appurtenant structures, remote Sensing and Geodetic Monitoring are often used to monitor the overall behaviour of the dam and catchment areas. Geodetic monitoring techniques, such as Global Navigation Satellite Systems (GNSS) are employed to measure precise horizontal and vertical displacements of the dam and its foundations, as well as of any potential unstable area such as landslides. Additional instrumentation such as inclinometers and piezometers can also be put in place where landslides have been identified.

## 3 GEOLOGY AND HYDROGEOLOGY OF FOUNDATIONS

The foundations and abutments of a dam are crucial components that play a significant role in the stability and safety of the structure. Several geological aspects need to be considered during the design and construction of the structure.

### 3.1 *Geological characteristics of the foundations*

The rock-matrix's composition and degree of weathering are essential factors to consider since the dam foundations need to support the weight and load distribution. Indeed, different rock types have distinct mechanical and physical properties. The composition, grain size, and mineralogy of the rock affect its strength, permeability, and durability, which are crucial geotechnical parameters. The weathering and alteration will also play a significant role in the quality of the rock mass. Generally, the weathered material will be removed from the foundation, but specific mitigation measures must be taken in case of localised deep-seated alteration.

In addition to the rock matrix, discontinuities, such as joints, faults or bedding planes, can have a significant impact on the geomechanical properties and stability of dam abutments and foundations. Discontinuities are natural planes of weakness in the rock mass, and their orientation determines how they interact with the applied loads and the overall stability of the dam structure. Analysis of discontinuities including their orientation, roughness, alteration, persistence, spacing and infill is essential to assess the geotechnical parameters to take into account in dam stability calculations.

An example of a potential problem due to discontinuities is shown in Figure 5. During the course of a geological study carried out as part of the heightening by 20 m of a dam in Africa, a specific study was performed with regards to the potential risk of sliding of some part of the rock mass due to the thrust from the dam and water pressure. The analysis requested a detailed assessment of the joint properties, in particular a subhorizontal fault system parallel to the bedding plane with low shear resistance. The result was then used to design a satisfactory dam upgrade.

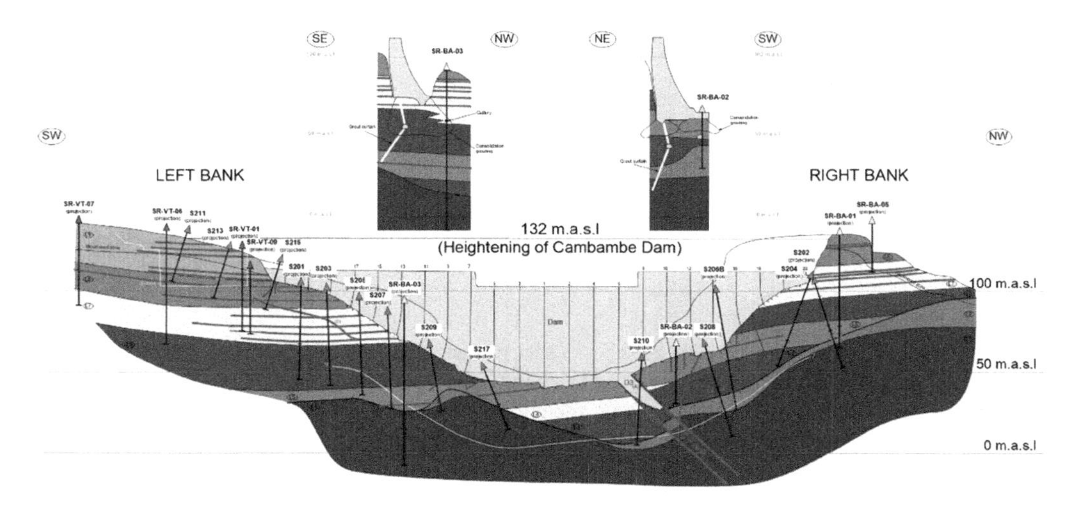


Figure 5. Geological cross-sections of the dam foundation and illustration of subhorizontal faults (red lines) in both abutments which present unfavourable conditions in combination with vertical joint sets - Africa.

It is not only the quality of the foundation that is important, but also the homogeneity of its geotechnical characteristics. The rock mass heterogeneity can significantly influence the behaviour of dam foundations, affecting factors such as shear strength, permeability and deformation patterns. This can be illustrated by the example of Les Toules dam in Switzerland, where the permanent deformations were significantly higher on the left bank, indicating that the rock mass was almost twice as deformable as on the right bank. This was due to the presence of a rock compartment on the left bank with geomechanical characteristics significantly inferior to those of the remaining of the foundation (Figure 6). This meant that requirements regarding seismic safety could not be met. Consequently, significant works had to be undertaken to reinforce the dam.

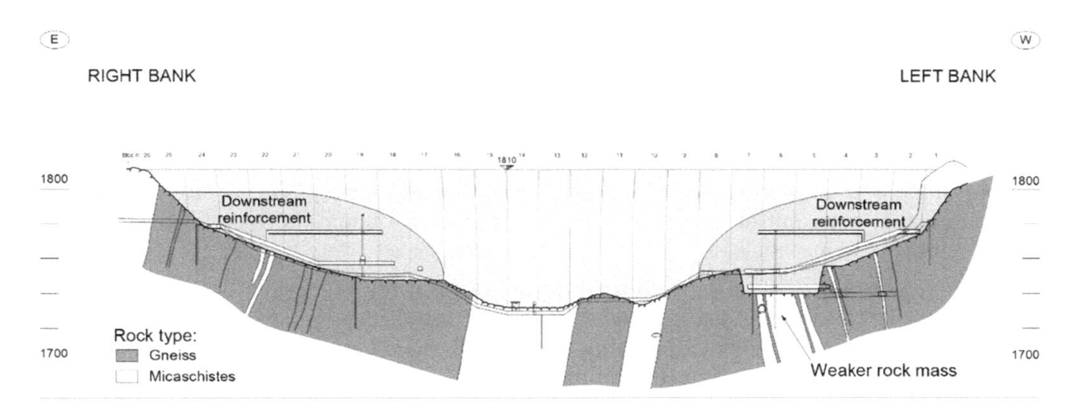


Figure 6. Les Toules, Switzerland - Geological cross section along the dam showing the heterogeneity of the foundations.

Another aspect where the rock mass quality plays an important role is regarding scouring which refers to the erosion and removal of rock caused by the flow of water. Scouring generally occurs during extreme events such as floods. It is a significant concern as it can undermine the stability of the dam or some dam organs. Factors influencing the scouring include the channel morphology, the water flow conditions, the reservoir sedimentation and the sediment characteristics and, of course, the rock mass properties which can be characterised by an erodibility index (Annandale, 1995).

A famous example is the case of Kariba dam in Zambia-Zimbabwe where an 85 m deep plunge pool scouring developed immediately downstream of the dam due to prolonged spillages during the first 20 years after construction (Michael, 2006). The chosen treatment measure was to reshape the plunge pool to avoid further scouring (Figure 7). The ongoing mitigation work requested the preliminary construction of a downstream cofferdam to isolate the reshaping works.

Figure 7. Kariba Dam, Zambia/Zimbabwe - View of the plunge pool (March 2023) and computer-generated view of the reshaped excavation (Razel Bec, 2016).

### 3.2 *Water pressure and possible seepage through the foundation*

High water pressure below the dam and/or seepage through the dam foundation are very sensitive issues since they could lead to unfavourable conditions in terms of dam safety in addition to the potential negative impact with regards to the loss of water. The well-known dam failures of (1) Malpasset dam in 1959 and (2) Teton dam in 1976 respectively illustrate these two problematics related to the hydraulic conditions within the foundation:

1. High hydraulic head combined with unfavourable orientations of joint sets may lead to unstable wedges below the dam or within its abutments. It is therefore crucial to identify and to assess the joint sets in order to determine if unfavourable geological structures prevail, and to properly design a grout curtain and drainage system to prevent unacceptable load conditions. If necessary, additional treatment of the upstream toe of the dam can be carried out.
2. Concentrated leaks within the rock mass at the base of the core may lead to the development of a continuous conduit at the base of the core and unacceptable erosion conditions. A systematic treatment of joints, an appropriate cut-off foundation (including filters) and appropriate core materials (low erodibility) are therefore essential.

Figure 8 illustrates some seepage located downstream of the Hongrin North Dam, on the right abutment. The water discharge was low but the installed piezometers confirmed water pressure up to 4 bars. Stability calculations of the rock mass have been performed and indicated that a significant improvement in term of safety factor can be achieved via a proper drainage of the rock mass, i.e. a reduction of the water pressure (Koliji et al, 2011a and 2011b; Bussard et al, 2015). A targeted and phased action plan has been implemented (Leroy et al, 2016) and rehabilitation works of the local defect of the grout curtain as well as additional drainage have been carried out in 2018 (Bussard & Wohnlich, 2018). The new conditions led to the drying up of the water outflow illustrated on Figure 10. The results indicate a significant decrease of the hydraulic head within the downstream right bank (general decrease of 70%) and a similar decrease of the seepage water discharge (decrease of 70%) which confirm a satisfying improvement in terms of slope stability.

Figure 8. Hongrin dams, Switzerland – The seepage that developed within the right abutment immediately after the reservoir impounding has been treated in 2018.

A numerical model is recommended to assess the likely hydraulic head distribution below the dam and potential seepage conditions. The model will assist in the appropriate design of the grout curtain and drainage system, as well as the installation of an adapted monitoring system.

Finally, the presence of soluble rocks within the dam foundation could be a critical issue, particularly in the case of extremely to highly soluble rock (halite, anhydrite and gypsum), which may evolve significantly during the life span of the dam (active development of karsts). Such types of rock should be clearly identified during the investigation phases in order to precisely determine the site conditions and if appropriate mitigation measures can be implemented or not. In case of less soluble rocks (carbonate: limestone, dolomite), there is a risk that existing cavities and conduits prevail within the foundation and abutments and could lead to potential seepage, including a risk of wash-out of infilling materials during the reservoir impounding.

## 4 GEOLOGY OF THE RESERVOIR AND CATCHMENT AREA

The geology of a dam reservoir and its catchment area plays a significant role not only regarding water leakage and therefore the long-term economic viability of the dam and its potential for sedimentation, but also in terms of the integrity and stability of the structure which can be threatened by natural hazards.

### 4.1 *Reservoir watertightness*

The presence of permeable features within the reservoir contour may lead to some leakage if they are connected to possible resurgence points located at a lower elevation than the reservoir water level.

The permeable features may consist of permeable formations, in particular soluble rocks or coarse and porous loose deposits, and possibly fault zones. The assessment of the risk of leakage requests a detailed analysis of the local and regional hydrogeological conditions. The implementation of mitigation measures could be an option (Figure 9) and a dedicated surveillance plan is recommended along the sensitive sections.

Figure 9. Example of a reservoir slope reshaping and a geomembrane lining installation for preventing water leakage in karst and fissured rock formations (Middle East).

### 4.2 *Natural hazards*

Dams are often located in mountainous areas which are prone to natural hazards such as landslides, debris flows and avalanches. If close to the structure, these hazards can damage the dam itself or essential dam structures, such as spillways, or compromise their functionality which can lead to overtopping of the dam.

Furthermore, the impact of natural hazards on a reservoir even far from the dam also needs to be addressed as it can have dramatic consequences. A famous example is the Vajont dam disaster in Italy which happened on 9 October 1963. The disaster was triggered by a massive landslide on Monte Toc, a mountain adjacent to the dam. An estimated 270 million cubic metres of rock and soil fell into the reservoir at high-speed creating a large wave that overtopped the dam. The Vajont Dam disaster is a tragic reminder of the potential impact of landslides on dam safety and the importance of thorough geological assessments, monitoring systems and emergency preparedness in areas prone to such hazards.

A sound assessment of the hazards around the reservoir must therefore be carried out by means of desk and in situ studies, analysis and computer modelling using state of the art softwares. Studies should encompass hazard identification and characterization, vulnerability assessment and risk analysis (Figure 10). Recommendations regarding mitigation strategies, monitoring and early warning systems should be made so that dam owners can effectively assess and manage natural hazards around the dam reservoir, therefore safeguarding the structure, downstream communities and the environment from potential risks.

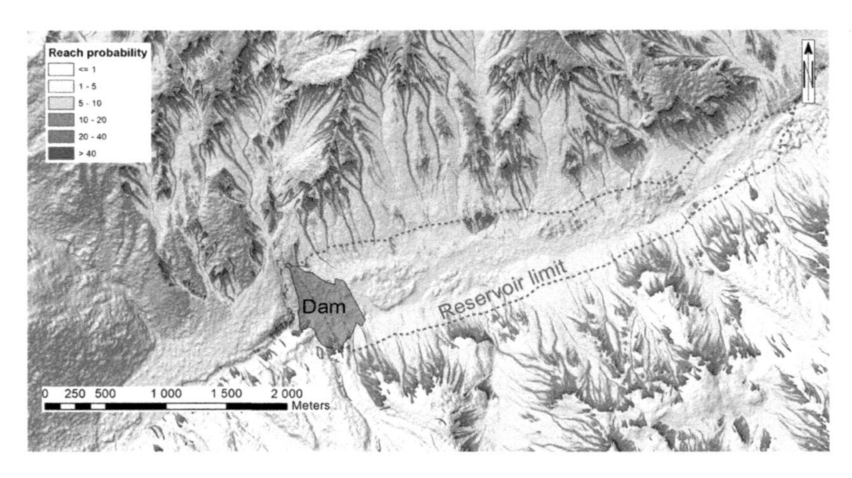


Figure 10. Example of an assessment of rock fall reach probability for a dam's reservoir project in Georgia.

With potentially the same tragic consequences as large landslides, Glacial Lake Outburst Floods (GLOFs) can have significant impacts on dam safety. GLOFs occur when a glacial lake dammed by a glacier or moraine fails, releasing a large volume of water downstream. These events can result in catastrophic flooding, with devastating consequences for dams located downstream. In the past few decades, the progressively warming climate has caused the volume of the glacial lakes to expand rapidly over the world increasing the risk. If a dam is in the path of the floodwaters, it may be overwhelmed by the excessive flow, potentially leading to structural damage or even complete failure.

In addition, the high-velocity floodwaters associated with a GLOF can cause erosion and scouring of the dam's foundation and surrounding areas. This erosion can weaken the dam's structure, compromise its stability, and undermine its foundation. Furthermore, GLOFs can transport substantial amounts of debris, including trees, rocks, and ice. This debris can accumulate at the dam site, potentially blocking water flow and obstructing spillways.

To mitigate the hazards posed by GLOFs, various measures such as remote sensing, early warning systems, monitoring of glacial lakes, reinforced dam designs, and emergency preparedness plans must be implemented (Weicai, 2022).

Climate Change is also degrading the permafrost conditions in mountainous areas. Such effects are already visible and have notable impacts in the Himalayan mountain belt. For instance, a debris flow of around 4 mio m3 occurred in Barcem in 2015 (Pamir mountain, Tajikistan) during heat wave conditions. The deposits dammed the Gunt valley and the upstream area was flooded almost to the Pamir 1 HPP. Reflection is in progress for building a derivation structure (bypass) for preventing a potential flooding of the powerplant during a next event. Disaster Risk Reduction and resilience strategy are necessary in the present changing environment.

## 5 CONCLUSIONS

In conclusion, geological studies play a crucial role in ensuring the safety and integrity of dams. The complexity of geological conditions at dam sites necessitates a thorough understanding of the underlying rock and/or soil properties, structural geology, and potential geological hazards.

Geological studies enable the selection of suitable dam sites, taking into account factors such as foundation conditions, seepage control and natural hazards. They also provide valuable insights for the design of dam foundations and structures considering the strength, permeability and deformation characteristics of the underlying geology.

Ultimately, the importance of geological studies for dam safety cannot be overstated. By integrating geological data by means of a geological model into the design, construction, and ongoing monitoring of dams, informed decisions can be made to mitigate risks.

By integrating geological knowledge and expertise into dam engineering practices, we can ensure the long-term safety.

## ACKNOWLEDGMENTS

The authors would like to thank the Swiss Dam Owners for their trust and continued collaboration over the years, and Gruner Stucky for their frequent requests for our geological expertise on projects around the world.

## REFERENCES

Annadale, G.W. 1995. Erodibility, Journal of Hydraulic Research, 33:4, 471–494.

Bussard, T., Wohnlich, A., Leroy, R. 2015. Monitoring of the Hongrin arch dams in Switzerland, water leakage and stability of the foundation issues. Colloque CFBR (Comité Français des Barrages et Réservoirs): Fondations des barrages: caractérisation, traitements, surveillance, réhabilitation. 8-9 April 2015. Chambéry, France.

Bussard, T., Wohnlich, A. 2018. Ouvrages de l'Hongrin. Travaux d'entretien 2018. Comité Suisse des Barrages. Interaction entre barrages et foundation. 20 August 2018. Montreux, Switzerland

Droz, P., Wohnlich, A. 2019. Rehabilitation of the monitoring system of Inga 1 and 2 dams, *Hydropower & Dams, Issue 2*, 2019

Fern, I., Jonneret, A., Kolly, J.C. 2019, Travaux de confortement de l'appui rive gauche du barrage de Lessoc, Wasser Energie Luft – 111. Jahrgang, 2019, Heft 3.

ICOLD. 1995. Dam Failure Statistical Analysis. Bulletin 99

Koliji, A., Bussard, T., Wohnlich, A., Leroy, R. 2011. Abutment stability assessment at the Hongrin arch dam. Hydropower & Dams. Issue Three, 2011.

Koliji, A., Bussard, T., Wohnlich, A., Zhao, J. 2011. Abutment stability assessment at the Hongrin arch dam using 3D distinct element method. 12th ISRM Congress. 16-21 October 2011. Beijing, China.

Leroy, R., Bussard, T., Wohnlich, A. 2016. Hongrin Dam – Long Term Serviceability of the Right Bank. HYDRO 2016. International Conference and exhibition. 10-12 October 2016. Montreux, Switzerland

Michael, F.G., Annandale, G.W., 2006. Kariba Dam Plunge Pool Scour, International Conference on Scour & Erosion, November 2006.

Razel Bec, 2016, Kariba Dam, Plunge pool reshaping, method statement (not published)

Swiss Federal Office of Energy SFOE. 2015. Directive on the Safety of Water Retaining Facilities

Weicai W., Taigang Z., Tandong Y., Baosheng A., 2022, Monitoring and early warning system of Cirenmaco glacial lake in the central Himalayas, International Journal of Disaster Risk Reduction, Volume 73, 15 April 2022, 102914.

World Bank. 2021. Good Practice Note on Dam Safety – Technical Note 2: Geotechnical Risk

*Role of Dams and Reservoirs in a Successful Energy Transition – Boes, Droz & Leroy (Eds)*


# Dams and photovoltaic plants – The Swiss experience

E. Rossetti, D. Maggetti & A. Balestra
*Lombardi Engineering Ltd., Switzerland*

ABSTRACT: Nowadays, energy transition is an increasingly frequent theme addressed by many actors belonging to a variety of sectors, from private institutions to governmental ones. Commonly, the main objective is to promote the transition from fossil fuels to renewable energy sources. This challenge is addressed in many ways, also depending on the administrative boundaries considered, and range from more local solutions to continental and global ones.

For example, in Switzerland, in 2017, the Swiss electorate accepted the revised Federal Energy Act (known as Energy Strategy 2050). This amendment promotes the energy transition, setting the basis for increasing the sustainability and efficiency of energy generation.

In this context, exploiting solar energy as a renewable resource plays an important role. However, the generic topographic limitations can represent an important restriction for the spreading of such installations. With a production pattern that fits well together with the flexibility of hydropower, innovative solutions, such as the possibility of placing photovoltaic (PV) panels on the downstream face of dams (dam mounted photovoltaics, DMPV) or over large reservoirs (floating photovoltaic, FPV), can represent a turning point to promote solar energy production.

In this article, such pioneering solutions are exploited, with a special look at the Swiss Alpine area. The authors also reported four examples of PV panels installation in such geographic region, which emphasizes the importance and the potential of these solutions to promote the diffusion of solar energy production.

## 1 INTRODUCTION

Switzerland, in 2017, adopted the new energy law known as the Energy Strategy 2050 (in the following ES2050). This law imposes a progressive transition from nuclear and fossil fuels by promoting renewable energy sources. Therefore, the question arises of replacing these energy vectors by building new solar, wind and hydroelectric plants and exploiting any other renewable production methods.

These include the possibility of building photovoltaic systems on almost any kind of surface, ranging from simple roofs and house facades to large industrial areas, car parks, motorways, and Alpine solar fields. The goals of the ES2050 are very ambitious and stimulate technological progress and research. For photovoltaics, the aim is to increase from the current 2.9 TWh/y of energy produced (2021) to a future 34 TWh/y. Photovoltaic installations are growing in number at a rapid pace, however, large areas are required to achieve the goals of the ES2050. Some good opportunities can be found at altitude, where certain advantageous aspects can positively influence photovoltaic production. A solar module in the mountains is expected to produce more than the same one in the lowlands and, above all, transfers most of its production to the winter (even more than +50% compared to the lowlands), which is the time of year when electricity is most required; these offsets can be reduced with Alpine photovoltaic systems. This positive effect is due to reduced blurring of the air (clouds and fog), a less filtering atmosphere, higher efficiency of the panels at low temperatures and the Albedo effect. The latter consists of the increased exposure of the panels due to the reflection of snow, which increases the amount of solar energy reaching the surface covered by the installations.

## 2 DMPV AND FPV INSTALLATIONS

This article deals with solar energy by discussing two types of photovoltaic systems with a particular focus on the Swiss context. The first concerns the installation of panels on dam walls (Dam Mounted Photovoltaics, DMPV), while the second involves the construction of floating platforms on Alpine lakes (Floating Photovoltaics, FPV).

As far as the DMPV installations are concerned, the favourable aspects are mainly the simplicity of the installation on a concrete structure and the reduced environmental impact, by taking advantage of an already anthropogenic and available area. On the other hand, there are some limitations, such as the exposure of the dam depending on its position and orientation, the slope of the downstream wall which is not very adaptable, and the shape of the dam (curvature).

Due to the larger possible plant area on the reservoir compared to the wall/dam surface, FPV plants have a larger potential energy production than DMPV plants. Furthermore, the alignment of FPV systems can be chosen at will and is not dictated by the dam. On the other hand, the impact of a DMPV on the landscape is lower and the installation effort is lower, which makes implementation easier.

## 3 POSSIBLE EVALUATION METHODOLOGY OF THE SWISS POTENTIAL

As part of two Master's theses at ETH Zurich, an evaluation method was developed and the potential and suitability of FPV and DMPV systems for 23 of the largest reservoirs and dams in Switzerland was examined (Rytz, 2020; Maddalena, 2021). Two evaluation matrices were developed, one for FPV and one for DMPV systems, which were divided into three main categories: "Acceptance", "Energy and Potential" and "Economics". the weighting was determined at the discretion of the authors and was based on potential studies on hydropower expansion (Ehrbar et al., 2019; Felix et al., 2020). The estimated potential production for the studied dams and reservoirs, ranges for FPVs from 350 to 450 GWh/y and for DMPVs from 11.5 to 14.5 GWh/y. Applying the method to all storage facilities in Switzerland, it is estimated that the potential could be between 500-1000 GWh/a for FPV and 15-20 GWh/a for DMPV (Maddalena et al, 2022).

## 4 CASE STUDIES

### 4.1 *FPV installation on the Lac des Toules*

On the Lac des Toules, in the canton of Valais, a prototype Alpine FPV installation was placed in 2019. The platform, located at 1810 m a.s.l., consists of 35 floating elements carrying 2'240 $m^2$ array of bifacial panels and a resulting installed capacity of some 426 kWp, expected to produce about 800 MWh/y (Romande Energie, 2019).

Purpose of this installation, which precedes the large-scale installation, was to verify the technical and financial feasibility of the project. Pilot tests between 2013 and 2019 suggested a 50% increase in power generation compared to plateau levels. From this data, the annual yield was estimated at 1'800 kWh per installed kWp. The actual result averaged 1'400 kWh, representing an increment of only 30% (Romande Energie, 2023). This difference can be explained by the fact that the floating plant was designed to maximize winter production and greater panel slope at the expense of the total energy production; for technical reasons it was also located further south and therefore closer to the mountains than the ground-based structure tested during the feasibility studies, resulting in additional shade, leading to up to one hour of lost sunshine exposure per day. In addition, snow – and especially drifting snow – caused a few days of downtime each year and damaged approximately ten of the photovoltaic (PV) panels.

The prototype is currently being adapted to prevent snow accumulation by installation of windbreakers. Several technical challenges have also been faced: it has been necessary to anchor the FPV to the bottom of the reservoir, thus allowing the structure of being theoretically able to withstand wind gusts of up to 120 km/h, ice layers of up to 60 cm and temperatures from -25 to 30 °C.

The experience gained during the prototype phase will enable Romande Energie to develop a more efficient large-scale installation. In the future, this plant is expected to be enlarged; indeed, developer's ambition is to add floating elements to reach a production of about 22 GWh/y (which should be the maximum potential available for that reservoir, i.e., enough for 6′200 households).

Figure 1. Pilot FPV plant on Lac des Toules (© Romande Energie).

### 4.2 *DMPV installation on the Muttsee dam*

On the Muttsee dam, in the canton of Glarus, a pioneering DMPV plant with a total power output of 2.2 MW has been installed between 2021 and 2022 by the companies Axpo and IWB. Its expected production of approximately 3.3 GWh/y (the annual demand of 950 households). This amount of annual energy is produced at 2500 m a.s.l. by about 5000 panels mounted on a dam surface of about 10'000 $m^2$. Approximately 50% of the production is generated in winter due to the favorable conditions caused by the altitude. Investment costs are about CHF 8 million (Heierli, 2022; Maggetti et al, 2023) and are mainly referable to the high logistic costs.

Figure 2. DMPV plant on the Muttsee dam (© AlpinSolar).

### 4.3 *DMPV installation on the Albigna dam*

Another example can be found in Albigna, in the canton of Grisons, where ewz (Elektrizitätswerk der Stadt Zürich) installed in summer 2020 a DMPV plant on 670 m of the dam's upstream face for 410 kWp capacity. The plant, located at 2165 m a.s.l., exploits the reflection of the sun's rays on the lake (in addition to the direct incidence) to amplify the energy production and manages to balance the energy between summer and winter, which in total is worth about 500 MWh/y (the annual needs of 210 households). Investment cost are about CHF 700'000 (ewz, 2020).

During the construction and operation of the plant, various synergies were taken in account:

- The grid connection at the Albigna dam was already in place.
- Most of the installation work was carried out by ewz employees, who have already initiated the pilot project.
- The year-round availability of own staff also simplifies any maintenance work.

Figure 3. DMPV plant on the Albigna dam (© ewz).

### 4.4 *DMPV installation on the Valle di Lei dam*

Very similar to the Albigna plant is the one on Valle di Lei arch dam in Grisons, also from ewz. Put in service in September 2022, the plant, which is also located on the upstream crest of the dam, is 550 m long and lies at an altitude of slightly less than 2000 m a.s.l., allowing it to have an increased efficiency of around 25% compared to the Plateau. The installed power of the plant is 343 kWp for a production of about 380 MWh/y (annual consumption of 110 households). Investment cost are about CHF 800'000 (ewz, 2022).

Figure 4. DMPV plant on the Valle di Lei dam (© ewz).

## 5 IMPLICATIONS FOR DAM SAFETY – THE MUTTSEE EXPERIENCE

With regard to the safety of the Muttsee dam, the supervisory authority (dams section, SFOE) formulated several requirements that had to be fulfilled:

- Stability: Verification of stability due to the changes in concrete temperatures caused by the placement of the solar modules over the entire area. In fact, the dam wall experiences a different temperature distribution due to the covering with solar panels. With an FE analysis, it could be proven that this temperature changes only result in a very insignificant change in the behaviour of the dam in the form of additional valley-side deformations of < 2mm. These minor, additional deformations do not affect the stability of the dam.
- Flood safety: Verification of flood safety by installing the solar modules in the spillway. The generously selected distance of the solar panels from the spillway crest (approx. 7 m) ensures that the function of the spillway is always guaranteed, and that no interference takes place due to the solar panels.
- Anchorages in the airside parament: Specification of the number and detailed design, the maximum forces introduced as well as the representation of the anchorage system of the supporting structure. With a specification of the anchorage used and an anchor depth that is not critical for the barrier, this verification could be provided without any problems.

- Visual inspection: Showing with a walk-on concept how the visual inspection can be carried out quickly and easily under the solar modules. A walkway between the airside parament and the solar panels (1.5 m distance) allows visual inspection of the concrete surface at any time. Furthermore, the individual block joints were kept free by a gap of 0.5 m between the panel blocks.
- Existing measuring equipment: Evidence that the monitoring of the Muttsee barrier is not affected by the construction or operation of the solar power plant. It was checked and proven that all visors for geodetic measurement remain free and that other measuring equipment is not restricted in its function by the PV plant.

## 6 OVERVIEW AND OUTLOOK

### 6.1 *Overview*

In the following Table 1, an overview of the by-now existing Swiss FPV/DMPV-plants is given and comparison is made with a reference Swiss Plateau rooftop installation in the city of Zürich computed with an internet-based tool (EnergieSchweiz, 2022).

Table 1. Overview of Swiss FPV and DMPV plants.

| PV-Plant name | Les Toules | AlpinSolar Muttsee | Albigna Solar | Lago di Lei Solar | Reference PV-plant in Zürich |
|---|---|---|---|---|---|
| Type | FPV | DMPV | DMPV | DMPV | Rooftop |
| Commissioning | 2019 | 2022 | 2020 | 2022 | - |
| Elevation [m asl] | 1'810 | 2'474 | 2'165 | 2'000 | 500 |
| Surface [$m^2$] | 2'240 | 10'000 | | | |
| Capacity [kWp] | About 400 | 2'184 | 410 | 343 | 150 |
| Expected annual generation [MWh/yr] | 530 | 3'300 | 500 | 380 | 150 |
| Specific yield [kWh/kWp] | 1'400 | 1'500 | 1'200 | 1'100 | 1'000 |
| Investment cost [million CHF] | Pilot project, not directly comparable | About 8 | 0.7 | 0.8 | 0.2 |
| Specific cost [CHF/Wp] | | 3.70 | 1.70 | 2.30 | 1.40 |

As a comparison, the average worldwide total investment cost of an FPV system in 2018 varied between US$0.8/Wp and 1.2/Wp, depending on the system's size and location. The CAPEX of large-scale but relatively uncomplicated FPV projects (around 50 MWp) was in the range of US$0.7-$0.8/Wp in the third and fourth quarters of 2018, depending on the location and the type of modules involved (World Bank Group, ESMAP and SERIS. 2019).

### 6.2 *Outlook*

The ES2050 targets of 34 TWh/a from PV remain ambitious, especially because in 2021, photovoltaics in Switzerland had an installed capacity of about 3.6 GW that produced 2.9 TWh. This would therefore be more than a tenfold increase of current production in 29 years. However, a very positive trend for the photovoltaic sector can be seen from 2019 onwards, as each year the installed capacity increased by around 40% compared to the previous year (+43% between 2020 and 2021).

The ES2050 poses several technological challenges, which can only be met with interdisciplinary solutions. Among these, solar energy will play a key role, and installations in Alpine contexts are confirmed to be interesting and increasingly competitive. The installation of photovoltaic panels on dams and in their corresponding reservoirs is a solution to be considered, especially because of the reduced impact on the landscape, the speed of installation and deployment (especially for DMPVs) and the possibility of balancing production between summer and winter.

DMPVs are likely to be the exception, but hydropower plants have considerable potential for new alpine PV plants due to their existing infrastructure.

To assess the suitability of the two discussed PV variants in Switzerland, a methodology has been proposed and the potential was assessed. With in-depth multi-criteria analyses, one should recognize the best alternatives and proceed further.

Some swiss case histories exist and useful lessons can already be drawn for further installations.

## ACKNOWLEDGEMENTS

The authors acknowledge the inputs regarding Muttsee dam safety provided by Rico Senti from Axpo Power AG.

## REFERENCES

Ehrbar, D.; Schmocker, L.; Vetsch, D.; Boes, R. (2019). Wasserkraftpotenzial in Gletscherrückzugsgebieten der Schweiz. Wasser Energie Luft 111(4): 205–212.

EnergieSchweiz (2022). Simulation von Energiesystemen mit dem Tachion-Simulation-Framework, Benutzerdokumentation, Oktober 2022. Available at: https://www.energieschweiz.ch/tools/solarrechner/ (accessed: May 29th, 2023)

Foen Hrsg. (2022) Energy strategy 2050. Available at: https://www.uvek.admin.ch/uvek/en/home/energy/energy-strategy-2050.html (Accessed: May 26th, 2023).

Ewz (2020), Albigna Solar – Erste hochalpine Solar-Grossanlage, Factsheet, Available at: https://www.ewz.ch/de/ueber-ewz/newsroom/medienmittteilungen/Solar-Albigna-produziert-ab-September.html (accessed: May 29th, 2023)

Ewz (2022), Lago di Lei Solar – Hochalpine Solargrossanlage auf der Staumauer Valle di Lei, Factsheet, Available at: https://www.ewz.ch/de/ueber-ewz/newsroom/medienmittteilungen/LagodiLei-Solar.html (accessed: May 29th, 2023)

Felix, D., Müller-Hagmann, M., Boes, R. (2020) Ausbaupotential der bestehenden Speicherseen in der Schweiz; *Wasser, Energie, Luft* 112(1): 1–10.

Heierli, C., 2022, A new Swiss partnership. *Waterpower Magazine*, November 2022: 16–17

Romande Energie (2019). Le premier parc solaire flottant en milieu alpin est en service; Communiqué de presse du 16.12.2019, Available at: https://www.solaireflottant-lestoules.ch/ (accessed, May 29th, 2023)

Romande Energie (2023). First ever high-altitude solar farm delivers initial findings; Press release 27.04.2023, Available at: https://www.solaireflottant-lestoules.ch/ (accessed, May 29th, 2023)

Rytz, S. (2021). Photovoltaics and hydropower reservoirs in Switzerland – Synergies and potential. *Master Thesis*, Laboratory of Hydraulics, Hydrology and Glaciology (VAW), ETH Zürich.

Maddalena, G., Hohermuth, B., Evers, F.M., Boes, R., Kahl, A., (2022). Photovoltaik und Wasserkraftspeicher in der Schweiz-Synergien und Potenzial. *Wasser, Energie, Luft* 114(3): 153–160 (in German).

Maggetti, D., Maugliani, F., Korell, A. Balestra, A., (2023). Photovoltaic on dams – Engineering challenges. *Proc. ICOLD European Club Symposium "Role of dams and reservoirs in a successful energy transition"* (Boes, R.M., Droz, P. & Leroy, R., eds.), Taylor & Francis, London.

Rytz, S. (2021) Photovoltaics and hydropower reservoirs in Switzerland – Synergies and potential. Master Thesis, Laboratory of Hydraulics, Hydrology and Glaciology (VAW), ETH Zürich.

World Bank Group, ESMAP and SERIS. 2019. Where Sun Meets Water: Floating Solar Handbook for Practitioners. Washington, DC: World Bank. Available at: https://www.worldbank.org/en/topic/energy/publication/where-sun-meets-water (accessed, May 29th, 2023)

*Role of Dams and Reservoirs in a Successful Energy Transition – Boes, Droz & Leroy (Eds)*
*© 2023 The Author(s), ISBN 978-1-032-57668-8*

# The importance of young professionals for dam engineering in Switzerland

Samuel Vorlet
*Hydraulic Constructions Platform (PL-LCH), Ecole Polytechnique Fédérale de Lausanne (EPFL), Switzerland*

Valentina Favero
*Dam supervision section, Swiss Federal Office of Energy (SFOE), Switzerland*

ABSTRACT: In Switzerland, hydropower is the main energy source and contributes to about 58.3% of the total production. The 2050 energy strategy aims to increase this share in the coming decades. This increase is a challenge for hydropower plants. The maintenance and rehabilitation of these facilities is therefore of paramount importance for their structural integrity and resilience, to guarantee a reliable and optimized operation in the medium and long term. Tomorrow's engineers will have a key role to play, and the great level of expertise acquired over the past decades in Switzerland must be maintained and transferred to the new generations of engineers. In fact, most of the dams in Switzerland have been constructed between the 50's and the 70's allowing a great level of expertise to be developed among dam engineers. The Young Professionals Group of the Swiss Committee on Dams, founded in 2019 and presented in this paper, has the role to ensure the knowledge transfer between generations.

## 1 INTRODUCTION

Decarbonization is a major challenge with a direct impact on energy strategies. It promotes the reduction of greenhouse gas emissions and the transition to renewable energy production in order to promote sustainability. In this context, hydropower is set to play a major role in the coming decades, due to its historical importance and its storage capacity, which enables operation reliability and compliance to complex grid requirements. In Switzerland, hydropower is the main source of energy, contributing to around 58.3% of the total domestic production (BFE 2022). The 2050 energy strategy aims to increase this share over the coming decades. This increase is a challenge for hydropower schemes, and maintenance and rehabilitation of these facilities are therefore of paramount importance for their structural integrity and resilience, to ensure reliable and optimized operation in the medium and long term. Tomorrow's engineers will therefore play a crucial role, and the level of expertise acquired over the decades in Switzerland must be maintained and passed on to the next generation of engineers. Indeed, most of dams in Switzerland were built between the 50s and the 70s, allowing a great level of expertise to be developed among dam engineers. This article briefly introduces the current situation of the next generation of dam engineers in Switzerland, presenting the importance of hydropower in Switzerland and the intergenerational transfer of knowledge. In addition, the recent creation of the Young Professionals group of the Swiss Dams Committee is presented as an example of how to encourage exchanges and knowledge transfer between different generations.

## 2 THE IMPORTANCE OF HYDROPOWER IN SWITZERLAND

Dams and hydraulic systems in Switzerland are an integral part of the country's history and geography. With most of its territory covered by the Alps, Switzerland is crossed by numerous rivers and has a dense network of Alpine reservoirs. In this favorable framework, hydropower has played an essential role in the country's energy, environmental and economic development. Over the course of the 20th century, Switzerland developed its considerable hydropower potential by building more than 200 large dams and numerous ancillary structures, enabling it to become a leader in hydropower production and contributing to its energy independence and sustainability.

Dams play an essential role in regulating water flows and inflows to optimize hydropower production. Reservoirs also manage and store water resources from precipitation and snowmelt, ensuring flood protection and drinking water supplies. Engineers have made an essential contribution to the design and optimization of these structures, and to the management of their operations. They have played an essential role in ensuring the proper operation of dams over the years, combining resource management with environmental protection.

Dam safety in Switzerland includes structural safety, monitoring, and maintenance, as well as an emergency plan to control the residual risk. The supervision of the structures is guaranteed by four levels of supervision including various levels of expertise (SDC 2015): the dam warden (level 1), an experienced professional (civil engineer) (level 2), confirmed experts (civil engineer and geologist) (level 3) and the supervisory authority (level 4). These different levels of supervision are essential to ensure optimum monitoring and maintenance of dams and reservoirs.

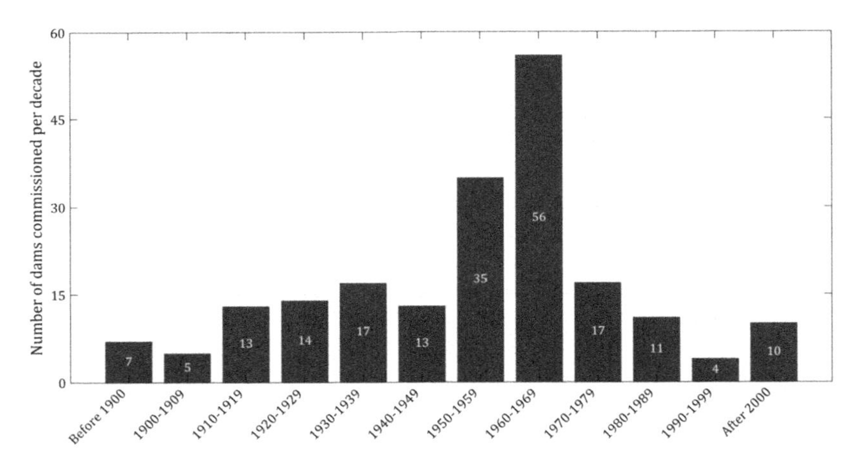


Figure 1. Number of dams commissioned per decade in Switzerland, according to Swiss Committee on Dams (SCD).

Although the construction of dams had already begun in Switzerland, it was between the 50s and 70s that almost half of all large dams were built, thanks to the favorable economic situation. During this period, more than 90 dams were built, and considerable experience was gained. This experience has led to the development of a high level of expertise in dam planification, design and construction, which is now internationally recognized. Yet, the number of dams commissioned in Switzerland strongly reduced in the following decades.

Today, experienced engineers often have the role of senior expert (level 3) and have an important wealth of technical knowledge. However, as most of them reach the end of their careers, it is essential to maintain an adequate level of expertise by investing in the education of future engineers and ensuring the transfer of knowledge to new generations, in order to guarantee the future competences.

## 3 YOUNG PROFESSIONALS GROUP OF THE SWISS COMMITTEE ON DAMS

The Young Professionals group of the Swiss Committee on Dams (SCD) was officially created at the General Assembly in Bern in 2019, with the aim of encouraging the participation and involvement of young professionals in the national committee, composed of Switzerland's leading experts in the field of dams, and promoting the transfer of knowledge to new generations of engineers. The creation of the group provides an opportunity for young professionals to meet and share experiences, and to promote the transfer of knowledge to the younger generation through meetings, seminars, conferences, technical visits, and exchanges organized with senior engineers and experts. It also promotes exchanges with other groups of young professionals in Europe and worldwide. SCD members under the age of 36 are members of the Young Professionals Group. Currently, the Young Professionals group has 42 members, including 10 women, from all over Switzerland. Members are mainly employed by engineering firms, operators, research and education institutions, or government.

The group meets officially four times a year. A core group of 4 to 5 people who meet monthly has also been set up to organize and manage the group on a more regular basis. The group is active within the SCD, since a representative of its members can participate in the various meetings of the main Working Groups, in order to promote knowledge transfer. Since its creation, the group has organized various activities in line with its objectives: various technical visits, guided by experienced engineers; visits to research facilities in Switzerland, enabling exchanges with industry professionals, in particular the Laboratory for Hydraulics, Hydrology and Glaciology (VAW) at the Swiss Federal Institute of Technology in Zurich (ETHZ); various meetings and conferences with recognized experts, notably in partnership with Hydro-Québec and the Swiss Federal Office of Energy (SFOE), among others. The group also works closely with the Young Professionals groups of other national committees, mainly in neighboring countries, and with the International Commission on Large Dams (ICOLD).

The creation of the Young Professionals group has made it possible, through the various activities carried out by the group, to foster the integration and participation of its members in the activities of the SCD, and to promote exchanges between Young Professionals and more experienced engineers, thereby promoting the transfer of knowledge. Passing on the experience gained by dam engineers to new generations in Switzerland is essential and is promoted by the group, which represents an example of how to encourage exchanges between people from the same profession, helping to prepare the next generation of engineers in the field of dams and hydraulics in Switzerland.

## 4 CONCLUSION

In a context of energy transition, the role of dams is essential to ensure a reliable and optimized energy supply in Switzerland in the medium and long term. Tomorrow's engineers will play an essential role in ensuring the operation, maintenance and rehabilitation of these structures, and the level of expertise acquired over the past decades in Switzerland needs to be passed on to new generations of engineers. This article briefly describes the current situation of the next generation of dam engineers in Switzerland, highlighting the importance of hydropower and intergenerational knowledge transfer in the dam sector. In addition, the recent creation of the Young Professionals group of the Swiss Dams Committee was presented as an example of how to encourage exchanges and knowledge transfer between different generations in Switzerland.

## REFERENCES

BFE 2022. Stand der Wasserkraftnutzung in der Schweiz am 31. Dezember 2022. *Bundesamt für Energie, Sektion Wasserkraft, Ittigen, Switzerland ; 2023 Apr. Report No.: 11375.*

SDC 2015. Role and duties of Dam Wardens. *Working Group on Dam observations. Wasser Energie Luft. 2015;107.*

*Theme A: Dams and reservoirs for hydropower*

*Role of Dams and Reservoirs in a Successful Energy Transition – Boes, Droz & Leroy (Eds)*


# How to win an international competition on sustainable sediment management

L. Gehrmann & T. Gross
*Hülskens Sediments GmbH, Wesel, Deutschland*

M. Detering
*D-Sediments GmbH, Werne, Deutschland*

ABSTRACT: The winning idea of the 2022 Guardians of the Reservoir Challenge, the Continuous Sediment Transfer, has shown great potential as an effective way to restore or maintain the functionality of reservoirs capacity and restore ecological conditions in downstream aquatic systems. The competition was organized by the Bureau of Reclamation (BoR) and United States Army Corps of Engineering (USACE) and had three stages: submitting an idea, developing a concept, and demonstrating the technology in a field test. The winning idea, which was selected by a technical jury, was Continuous Sediment Transfer, a near-natural method for autonomous restoring and maintaining reservoir volume and restoring sediment conditions in the tailwater.

The team's technology, the SediMover, was used to remobilize the sediment inside the reservoir and transfer it to the tailwater. Its modularity allows for adaptation to almost every situation, including difficult situations such as flood events and dry falls or quantitative limitations in sediment transfer. The team successfully demonstrated the process in a field test, which was presented in a demonstration video, a presentation, and a written essay. The technology is expected to undercut the cost of applying conventional dredging methods for reservoirs, thus saving reservoir operators money. Nevertheless, the new technology can be combined with conventional dredging technology and other equipment. After the pilot application in the competition, the team already has three larger commercial projects in preparation in Germany.

In addition to the practical need for application, the team aims for a broad and scalable application of the developed technologies. The judicious application of autonomous dredging will not only ensure cost-effective remediation of reservoirs and ensure their future operation but will also restore sediment continuity in rivers in a naturalistic manner. The benefits are not limited to the immediate downstream sections of the river but extend far downstream to river deltas and coastal regions, preventing further erosion damage. In this way, ecological benefits are achieved at no additional cost.

Moreover, the team is participating in developing guidelines for sediment management in reservoirs and researching the combination of sediment methane gas harvesting with the continuous sediment transfer, using different autonomous vessel types. The team is also working to adapt the technology to harvest methane generated inside the sediment to combat greenhouse gas emissions. Furthermore, a smaller version of the SediMover, the MiniMover, has been developed for use in tributaries, limited spaces of reservoirs or shoreline areas.

The Continuous Sediment Transfer process conducted with autonomous vessels shows great potential as it combines ecological benefits with economic benefits to combat a global problem. Reservoir operators and the government will have to invest significant sums in the coming years to ensure water supply in several regions of the USA and in Europe. Therefore, the automation achieved by the technology will ensure that it is able to undercut the cost of applying conventional dredging methods for reservoirs, thus saving reservoir operators money. The team's technology is expected to have good prospects as it offers both ecological and economic benefits.

# Specialist grouting works in the renewal of Ritom HPP, Switzerland

A. Heizmann
*Renesco GmbH, Lörrach, Germany*

G. Lilliu
*Renesco Holding AG, Chiasso, Switzerland*

ABSTRACT: This paper presents an approach to grouting in a highly water bearing ground, by means of a hybrid grout. A cement suspension is mixed with a variable ratio of polyurethane and depending on the pressure development during grouting, the percentage of cement suspension and polyurethane are correspondingly adjusted. Both the selection of the cement type, of the ratio water/cement and of the polyurethane system potentially allow for a wide range of application. The method was first applied in 2019, during excavation of the new headrace tunnel at Ritom HPP, in Switzerland. An unexpected strong water inflow in a highly fractured rock mass called for the selection of a suitable grouting technique to seal the rock without being washed out. Hybrid grouting proved to be very efficient in stopping the water inflow, and economical. Following the application at Ritom HPP, a laboratory test campaign was conducted to study the short- and long-term properties of the hybrid grout. The effect of adding polyurethane is to significantly reduce the setting time. It also affects some of the mechanical properties, e.g. compressive strength and Young's modulus. Additional laboratory testing aimed to optimize the grout design with respect to specific applications, and large-scale projects will help to refine and further improve the efficiency of the hybrid approach.

*Role of Dams and Reservoirs in a Successful Energy Transition – Boes, Droz & Leroy (Eds)*
*© 2023 The Author(s), ISBN 978-1-032-57668-8*

# Wave return walls within the adapted freeboard design at dams

Max Heß & Dirk Carstensen
*Institute of Hydraulic Engineering and Water Resources Management, Technische Hochschule Nürnberg Georg Simon Ohm, Nuremberg, Germany*

ABSTRACT: Optimizing freeboard at dams, became more and more relevant for operators of dams due to climate change, the assurance of flood retention, or the adjustment of the design water levels. The wave run-up represents the dominant portion of the freeboard at reservoirs and therefore deserves the most attention regarding optimizations. With the usage of wave deflection structures the safety of overtopping can significantly be increased. In contrast to coastal engineering overtopping as a result of the wave run-up with or without wave return walls in the adapted freeboard design at dams is generally not tolerated (zero overtopping). However existing design approaches for wave deflectors are currently always based on the overtopping rate, which therefore is greater than zero.

As part of independent investigations a hydrodynamic numerical model had been developed, calibrated and validated by using data obtained from physical hydraulic model tests. Based on performed simulations the general understanding of the redirection process of the wave run-up surge had been studied. Furthermore a data set, that specifies the required wave deflector size and the forces acting on the deflector at maximum (optimal) capacity utilization without overtopping was developed and used to build a dimensioning concept depending on the boundary conditions mentioned.

*Role of Dams and Reservoirs in a Successful Energy Transition – Boes, Droz & Leroy (Eds)*


# PV plant - New potentials in Vau i Dejes HPP, Albania

A. Jovani
*Albanian National Committee on Large Dams (ALBCOLD), Tirana, Albania*

F. Shaha
*Albanian Young Engineers Forum, Tirana, Albania*

ABSTRACT: This paper provides the information and results on the implementation of a photovoltaic (PV) plant on the Qyrsaq Dam of Vau i Dejes Hydropower plant (HPP) in Albania. Qyrsaq Dam is 54 meters high and has a dam crest length of 514 meters. This composite dam consists of a rock-fill dam with a clay core and a concrete gravity dam. The concrete gravity dam includes the power water intake and spillway systems, with a total capacity of 3900 $m^3/s$. The main purpose of the implementation of this project is the production of energy through the use of solar energy using the surface of the downstream face of Qyrsaq Dam. The project is an attempt to implement new technology in this field as well as the possibility of installing PV plants in other dams in the future. Albania is a country rich in water and sunlight, with a hydrographic catchment area covering 43700 $km^2$. Approximately 35% of the average water inflow is used for hydro energy. Moreover, Albania experiences over 300 sunny days per year. For 3 years, the Albanian Government has started the implementation of a strategy for installation of PV plants with capacity of 400 MW in some areas of our country. Till now, 3% of the energy is produced by them. Furthermore, the synthesis will give the requirements of ALBCOLD for installation of PV plant on large dams.

*Role of Dams and Reservoirs in a Successful Energy Transition – Boes, Droz & Leroy (Eds)*
*© 2023 The Author(s), ISBN 978-1-032-57668-8*

# Drini River Cascade - Unique in Europe

A. Jovani
*Albanian National Committee of Large Dams, Tirana, Albania*

E. Verdho, E. Kacurri & E. Qosja
*Albanian Power Corporation, Tirana, Albania*

ABSTRACT: This paper delves into the fascinating Drini River Cascade, a cascade of six large reservoirs and five Hydro power plants (HPPs), that serves Albania and North Macedonia. The Drini River Cascade is a unique and remarkable engineering feat in Europe, boasting a total water capacity of 4.3 billion m$^3$, and remarkable infrastructure. It features the tallest rockfill dam with clay core in Europe - the Fierza HPP dam, which stands at a staggering height of 166.5 m, and the largest Hydro power plant (HPP) with Hydro-matrix turbines-the Ashta 1 HPP and Ashta 2 HPP, with a total capacity of 53 MW.

Drini River flow from Ohrid Lake in Struga town of North Macedonia and Zhleb Mountain near of Peja town of Kosovo. They receives many relativity long tributaries. White Drin reaches the town of Kukes in Albania where it meet the Black Drin and forms the Drin River which flow into the Adriatic sea. At 335 km long, the Drin is the longest river of Albania of which 285 km passes across Albania and the remainder through Kosova and North Macedonia.

The Drini River Cascade's ability to harness the power of water for energy production is un-matched in Albania, with over 70% of the country's energy being generated by the cascade's HPPs. Moreover, the cascade's versatility is its strength, as it provides an array of services including hydro power production, water transport, aquaculture, tourism, flood protection, and even solar energy production.

The Drini River Cascade's future looks bright, with plans underway to construct a new HPP and expand its solar energy production capabilities.

In conclusion, this paper provides a comprehensive overview of the Drini River Cascade, highlighting its engineering marvels, remarkable energy production capabilities, and versatile applications, as well as its future prospects for dam safety improvement, expansion and new development.

*Role of Dams and Reservoirs in a Successful Energy Transition – Boes, Droz & Leroy (Eds)*


# Photovoltaics and hydropower – Potential study at Alpine reservoirs in Switzerland

G. Maddalena
*beffa tognacca, Claro, Switzerland*
*Laboratorium$^{3D}$, Biasca, Switzerland*

B. Hohermuth, F.M. Evers & R.M. Boes
*Laboratory of Hydraulics, Hydrology and Glaciology (VAW), ETH Zurich, Zürich, Switzerland*

A. Kahl
*SUNWELL, Lausanne, Switzerland*

ABSTRACT: Photovoltaics (PV) play a major role in Switzerland's energy transition. Compared to conventional installations in urbanized areas, PV at high altitudes may yield a 30% increase in power generation due to stronger solar irradiation and lower module temperatures as well as a considerably higher production share in winter. Numerous Swiss hydropower reservoirs are situated above 1,000m a.s.l. and already offer a good accessibility and grid connection. Therefore, they could provide an ideal base infrastructure for the installation of PV. The two main options for PV operation at reservoirs include modules mounted to the dam structure and modules floating on the reservoir.

This study investigates 23 selected Swiss hydropower reservoirs with 26 dams at elevations between 1,200 and 2,500m a.s.l. regarding their power production potential. The developed work-flow comprises a GIS analysis including inter-daily as well as seasonal shading effects. The resulting electricity production from solar radiation impinging on a PV module installed at a given orientation and tilt are simulated using the software SUNWELL. For validation, the simulated electricity production at selected sites is compared to planned or recently installed PV plants. The study ranks the investigated sites according to a multicriteria assessment matrix considering *societal acceptance*, *energy yield* and *economic feasibility*. The estimated total electricity production at the investigated sites amounts to 11.5 to 14.5 GWh/a for dam-mounted PV and 370 to 490 GWh/a for floating PV. The developed work-flow may also be applied to assess other promising PV installation sites.

*Role of Dams and Reservoirs in a Successful Energy Transition – Boes, Droz & Leroy (Eds)*
*© 2023 The Author(s), ISBN 978-1-032-57668-8*

# Photovoltaic on dams – Engineering challenges

D. Maggetti, F. Maugliani, A. Korell & A. Balestra
*Lombardi Engineering Ltd., Switzerland*

ABSTRACT: The installation of photovoltaic (PV) systems on dams is a pioneering initiative and an exciting interdisciplinary challenge.

The performance of a photovoltaic system strongly depends on its location, altitude, and exposure. Many existing dams offer a downstream surface of large extension suitable for the installation of a high-power PV plant; in the case of southward-oriented dams (located in the northern hemisphere) the conditions of incident solar radiation are optimal and therefore potentially capable of high efficiency.

The installation on existing structures of PV plants is particularly interesting due to the simplified authorization process: PV technology is broadly accepted in public opinion and in the case of installation on dams presents limited, or no dam safety issues.

A further advantage of these types of installations is the connection to the existing powerlines of the dam and to the nearby located hydropower plant, already connected to the distribution grid. No major additional works for the connection of the PV plant are thus required.

The design of PV plants on dams must match the characteristics of the existing structures as well as O&M requirements of PV plants: specific structural and access solutions must be developed.

Some of the authors were involved since the first ideas until the final commissioning in 2022 of the by-now largest Alpine PV power plant on the downstream face of Muttsee concrete dam, Glarus, Switzerland. At an elevation of 2'474 m a.s.l. and with a surface of 10'000 m$^2$, the plant shows a peak capacity of 2.2 MW with an expected annual yield of 3.3 GWh, mostly in Winter months.

RÉSUMÉ: L'installation de systèmes photovoltaïques (PV) sur les barrages est une initiative pionnière et un défi interdisciplinaire passionnant.

La performance d'un système photovoltaïque dépend fortement de son emplacement, de son altitude et de son exposition. De nombreux barrages existants offrent une surface aval de grande extension adaptée à l'installation d'une centrale photovoltaïque de grande puissance; dans le cas des barrages orientés vers le sud (situés dans l'hémisphère nord), les conditions de rayonnement solaire incident sont optimales et donc potentiellement capables d'un rendement élevé.

L'installation de centrales photovoltaïques sur des structures existantes est particulièrement intéressante grâce au processus d'autorisation simplifié. La technologie photovoltaïque est largement acceptée par l'opinion publique et, dans le cas d'une installation sur un barrage, elle ne pose que peu ou pas de problèmes de sécurité.

Un autre avantage de ce type d'installation est la connexion aux lignes électriques existantes du barrage et à la centrale hydroélectrique située à proximité, déjà connectée au réseau de distribution. Il n'est donc pas nécessaire d'effectuer des travaux supplémentaires importants pour le raccordement de l'installation photovoltaïque.

La conception des centrales photovoltaïques sur les barrages doit correspondre aux caractéristiques des structures existantes ainsi qu'aux exigences d'exploitation et de maintenance des centrales photovoltaïques: des solutions structurelles et d'accès spécifiques doivent être développées.

Certains des auteurs ont été impliqués depuis les premières idées jusqu'à la mise en service finale en 2022 de la plus grande centrale photovoltaïque alpine sur la face aval du barrage en béton de Muttsee, à Glaris, en Suisse. À une altitude de 2'474 m et avec une surface de 10'000 m$^2$, la centrale a une capacité de pointe de 2,2 MW avec un rendement annuel prévu de 3,3 GWh, principalement pendant les mois d'hiver.

*Role of Dams and Reservoirs in a Successful Energy Transition – Boes, Droz & Leroy (Eds)*


# Vianden pumped storage plant - large-scale shear tests on rockfill material of the upper reservoir ring-dam

Ch. Meyer & K. Thermann
*Tractebel Engineering, Bad Vilbel, Germany*

ABSTRACT: Located on the Our River on Luxembourg's eastern border with Germany, the Vianden pumped storage power plant was commissioned in 1964. The plant has a two-section upper basin, which was built on the summit plateau of the Nikolausberg mountain. The ring dam enclosing the upper basin is designed as a rockfill dam with an asphalt surface sealing. In the context of a routine safety review of the ring dam, in 2021 large-scale triaxial shear tests were carried out on the existing dam fill material. The aim of the tests was to verify the shear parameters adopted for the design considering an operating period of almost 60 years with regard to the stability calculations to be carried out. The paper describes the tests carried out and presents the results in the specific setting of the investigation program, which was performed as a basis for the safety review. The test results are also put into perspective of the frequently used approach for rockfill dams from Leps (1970) with a stress-dependent friction angle for different material qualities.

*Role of Dams and Reservoirs in a Successful Energy Transition – Boes, Droz & Leroy (Eds)*
 *ISBN 978-1-032-57668-8*

# Capacity building of dam wardens

A. Mico, N.-V. Bretz, O. Sarrasin & J. Fluixa-Sanmartin
*HYDRO Exploitation, Sion, Switzerland*

ABSTRACT: Dam wardens are responsible for carrying out measurements and visual inspections of dams. Their duties constitute the foundation of the four-level structure of Swiss dam surveillance. HYDRO Exploitation SA has been providing services in the operation and maintenance of hydropower plants since 2003. It employs approximately 40 dam wardens and periodically organizes technical training courses for them. These trainings offer a useful tool for new and experienced dam wardens to better apprehend their role in the dam's safety management. During these so-called capacity building events, dam wardens are also taught practical aspects on site, which represent an excellent opportunity to share experiences and improve their know-how.

# Collection and dissemination of knowledge on dams of Italy

S. Munari
*Enel Green Power – ITCOLD, Turin, Italy*

ABSTRACT: The knowledge documented in technical literature about a specific dam is a valuable source of information when dealing with the control and assessment of the dam. It is therefore essential to keep the knowledge acquired in the past. ITCOLD, through a specific working group, since several years has promoted the collection and dissemination of articles and memories concerning the dams in operation in Italy, through an ITCOLD Bulletin dedicated to this subject. A first edition of the Bulletin was published in 2012 up to the last update of 2019 in which the number of 2000 collected memories broke through. With the latest update of the bulletin, the reader's experience has been made more usable, making it easier to search and download the article of interest. In the coming years, ITCOLD will continue this path of disseminating knowledge on dams and in this sense the creation of an internal online library of indexes will only help this purpose. This article therefore intends to describe the results achieved by the ITCOLD working group in recent years and sets the objectives for the near future.

RÉSUMÉ: Les connaissances documentées dans la littérature technique sur un barrage spécifique sont une source précieuse d'informations lorsqu'il s'agit du contrôle et de l'évaluation du barrage. Il est donc essentiel de conserver les connaissances acquises dans le passé. ITCOLD, à travers un groupe de travail spécifique, promeut depuis plusieurs années la collecte et la diffusion d'articles et de mémoires concernant les barrages en exploitation en Italie, à travers un Bulletin ITCOLD dédié à ce sujet. Une première édition du Bulletin a été publiée en 2012 jusqu'à la dernière mise à jour de 2019 dans laquelle le nombre de 2000 documents techniques collectés a percé. Avec la dernière mise à jour du bulletin, l'expérience du lecteur a été rendue plus utilisable, ce qui facilite la recherche et le téléchargement de l'article qui l'intéresse. Dans les années à venir, ITCOLD poursuivra cette voie de diffusion des connaissances sur les barrages et en ce sens la création d'une bibliothèque interne en ligne d'index ne fera que contribuer à cet objectif. Cet article entend donc décrire les résultats obtenus par le groupe de travail ITCOLD ces dernières années et en fixe les objectifs pour le futur proche.

*Role of Dams and Reservoirs in a Successful Energy Transition – Boes, Droz & Leroy (Eds)*
*© 2023 The Author(s), ISBN 978-1-032-57668-8*

# Flood control across hydropower dams: The value of safety

Christina Ntemiroglou, Georgia-Konstantina Sakki & Andreas Efstratiadis
*Department of Water Resources and Environmental Engineering, National Technical University of Athens, Greece*

ABSTRACT: Hydropower reservoirs inherently serving as major flood protection infrastructures, are commonly occupied with gated spillways, to increase both their storage capacity and head. From an operational viewpoint, during severe flood events, this feature raises challenging conflicts with respect to combined management of turbines and gates. From the perception of safety, a fully conservative policy that aims to diminish the possibility of dam overtopping, imposes to operate the turbines in their maximum capacity and, simultaneously, opening the gates to allow uncontrolled flow over the spillway. Yet, this practice may have negative economic impacts from three aspects. First, significant amounts of water that could be stored for generating energy and also fulfilling other uses, are lost. Second, the activation of turbines may be in contrast with the associated hydropower scheduling (e.g., generation of firm energy only during peak hours, when the market value of electricity is high). Last, the flood wave through the spillway may cause unnecessary damages to downstream areas. In this vein, this fpaper aims to reveal the problem of ensuring a best-compromise equilibrium between the overall objective of maximizing the benefits from hydropower production and minimizing flood risk. In order to explore the multiple methodological and practical challenges from a real-world perspective, we take as example one of the largest hydroelectric dams of Greece, i.e., Pournari at Arachthos River, Epirus (useful storage 310 hm$^3$, power capacity 300 MW). Interestingly, this dam is located just upstream of the city of Arta, thus its control is absolutely crucial for about 25 000 residents. Based on historical flood events, as well as hypothetical floods (e.g., used within spillway design), we seek for a generic flood management policy, to fulfil the two aforementioned objectives. The proposed policy is contrasted with established rules and actual manipulations by the dam operators.

*Role of Dams and Reservoirs in a Successful Energy Transition – Boes, Droz & Leroy (Eds)*
*© 2023 The Author(s), ISBN 978-1-032-57668-8*

# Blasting excavation close to fresh Roller Compacted Concrete, in RCC Dam construction sites

P. Ruffato
*Civil Engineer, Webuild, Milano, Italy*

R. Folchi
*Nitrex Srl, Lonato, Brascia, Italy*

ABSTRACT: RCC dam construction is a long process which takes years to be completed, and requires precise planning and a tight sequence of activities. Due to planning constraints, RCC placement is to start before the completion of the dam foundation blasting excavation. The purpose of this paper is to demonstrate the possibility to perform blast excavation while in the same foundation area RCC is under placing at a distance in the order of 50m or less to the blasting, and compacted by vibrating rollers. The vibrations generated by rollers at different levels into the RCC were measured and compared with vibrations generated by the nearby excavations executed with explosive. It is assumed that controlled blasting excavation is safe in the vicinity of fresh RCC, if the induced vibrations are in the same order as those generated by rollers during the compaction operations. Laboratory tests on fresh RCC that sensed the reduced vibrations compared to those on RCC that did not, confirm that its material properties are not affected by the controlled blasting. The outcome of this work and added value on an RCC dam construction lies in allowing the excavation of a part of the dam foundations to be completed while RCC concrete placement is already in progress in the other part of the foundation.

*Role of Dams and Reservoirs in a Successful Energy Transition – Boes, Droz & Leroy (Eds)*
*© 2023 The Author(s), ISBN 978-1-032-57668-8*

# Swiss dam safety regulation: Framework, recent changes and future perspectives

M.V. Schwager, A. Askarinejad, B. Friedli, P.W. Oberender, A.J. Pachoud & L. Pfister
*Swiss Federal Office of Energy, Bern, Switzerland*

ABSTRACT: This paper outlines the current structure and evolution of the Swiss dam safety regulations and provides insights into the underlying reasoning and research. The structure of the Swiss regulatory framework is based on three pillars: structural safety requirements, surveillance and emergency planning. Emphasis is given to recent changes of the regulations due to ongoing advancements in the field of dam safety and related fields, reflecting the dynamic nature of the regulations. In particular, four different challenges are discussed. They relate to updating the requirements following the latest change of the Swiss national seismic hazard model, dealing with uncertainties regarding flood safety, preparing for future influence of dam ageing and accounting for particularities of natural hazard protection dams. Conclusions are drawn how to address the discussed challenges in the framework of potential future changes in the dam safety regulation.

*Role of Dams and Reservoirs in a Successful Energy Transition – Boes, Droz & Leroy (Eds)*
*© 2023 The Author(s), ISBN 978-1-032-57668-8*

# Châtelard basin storage expansion by making use of a spoil area

H. Stahl
*AFRY Switzerland Ltd., Zürich, Switzerland*

B. Müller & Ch. Hirt
*Swiss Federal Railways, Zollikofen, Switzerland*

B. Romero
*AFRY Switzerland Ltd., Lausanne, Switzerland*

ABSTRACT: The compensation reservoir Châtelard is an asphalt lined basin owned and operated by the Swiss Federal Railways. It is located in a narrow valley that has been partly filled to the downstream by a spoil area of tunnel muck derived from the construction of Nant de Drance pump storage scheme. The area between the asphalt lined embankment dam and the spoil area provided good opportunity to increase the storage volume of the basin. The Swiss Federal Railways took a brave decision for the alternative with the greatest increase of storage volume but also maximum impact on the existing scheme. With the chosen alternative, the active storage volume was doubled from 200'000 $m^3$ to 394'000 $m^3$.

The spoil area was not originally planned to serve as a water retaining structure and efforts needed to be made to investigate its properties and analyze its stability under static and dynamic loads as well as post-earthquake state with ruptured surface sealing. It was concluded that the original grade of the spoil area slopes of 1:1.5 (v:h) is too steep to meet the required factors of safety for a water retaining dam and an additional embankment to flatten the slope to 1:2.5 (v:h) was planned and implemented. In agreement with the supervising authorities, a fiber-optic deformation measurement system was included within the asphalt sealing of the new dam in order to monitor closely the embankment for settlements and the asphalt sealing for unacceptable strains during first impounding. The monitoring system was enhanced with geodetic measurements, piezometers within the embankment and a drainage system beneath the asphalt sealing.

The project implementation started in April 2021 and first impounding took place in February and March 2022. The first full impounding resulted in strains of 250 microstrains (0.025% or 0.25 mm on 1 m) in different areas of the new dam. Only along the plinth between the asphalt surface sealing and the rock cliff strains of up to 1'500 microstrains (1.5 mm on 1 m) were observed. All in all, the deformations observed with the fiber optic cables remained small and well distributed (no abrupt changes), indicating no deformation critical to watertightness. Three weeks after first full impounding, the geodetic measurements showed a maximum horizontal displacement of the five measuring points located on the new dam of 3.5 mm and a maximum settlement of 6.8 mm. After 9 months of operation, the horizontal deformation is still of the same order of magnitude and the settlements only increased slightly to 8.8 mm. These values confirm an appropriate compaction of the spoil deposit and the new earthworks embankment with flattened slope. Also, the piezometric pressures and drainage water quantities confirmed a satisfactory behavior during first impounding. After 9 months of normal operation the behavior is still fully up to expectation.

With this project, the Federal Railways facilitated a considerable increase in the efficiency of the existing scheme and improved its capability for flexible energy storage and just in time generation according to the ever-changing demands.

*Role of Dams and Reservoirs in a Successful Energy Transition – Boes, Droz & Leroy (Eds)*
*© 2023 The Author(s), ISBN 978-1-032-57668-8*

# An application of sophisticated FEM and simplified methods to the seismic response analysis of an asphalt-concrete core rockfill dam

A.D. Tzenkov
*Gruner Stucky Ltd, Renens, Switzerland*

D.S. Kisliakov
*University of Architecture, Civil Engineering and Geodesy, Sofia, Bulgaria*

M.V. Schwager
*Swiss Federal Office of Energy SFOE, Bern, Switzerland*

ABSTRACT: Despite the intensive development of sophisticated dynamic analysis methods and their implementation in the field of dam engineering over the last decades, some well-established simplified methods are still widely used for a rapid evaluation of the most important response parameters of an embankment dam under earthquake excitation. The present work deals with the application of a sophisticated FEM analysis and two simplified methods to the analysis of the seismic response of a rockfill dam with asphalt-concrete core (ACRD) in Bulgaria. The results of the two approaches are compared and conclusions are drawn as to the applicability of the simplified methods used for a typical ACRD.

*Role of Dams and Reservoirs in a Successful Energy Transition – Boes, Droz & Leroy (Eds)*


# Trift Arch Dam – an opportunity for hydropower generation due to a retreating glacier

A.D. Tzenkov, O. Vallotton & A. Mellal
*Gruner Stucky Ltd, Renens, Switzerland*

ABSTRACT: This article presents the concept for construction of an arch dam at the former downstream end of the retreating Trift Glacier in Switzerland. It describes the context of the project, the approach adopted for defining and optimising the shape of the arch dam, as well as the static and dynamic structural analyses carried out to verify the dam safety.

*Role of Dams and Reservoirs in a Successful Energy Transition – Boes, Droz & Leroy (Eds)*
*© 2023 The Author(s), ISBN 978-1-032-57668-8*

# Fulfilling pumped storage plants requirements with advanced geomembrane technology

G. Vaschetti & M. Scarella
*Carpi Tech, Balerna, Switzerland*

ABSTRACT: The construction of pumped storage plants is increasing due to the need of providing grid balance as wind and solar production increases. Pumped storage plants are demanding structures, due to the continuous variations of the water level that result in repeated loading and unloading conditions impacting on all the components of the plant. In particular, since the upper and lower reservoirs are often formed by excavating the natural slopes and by earthen or rock embankments, it is important to ensure that possible settlements and differential displacements will not affect the watertightness and will not cause uncontrolled water seepage. Waterproofing the reservoirs with conventional liners, such as concrete or bituminous concrete, has shown some drawbacks. The rigid liners demonstrated a poor capability to accommodate settlements, to provide efficient joint sealing, and to preserve the dimensional stability under temperature variations. As a consequence, the rigid liners often need maintenance, which in some cases implies outage of the plant and revenue losses. Geomembrane liners are characterised by outstanding endurance properties and are considered an efficient and durable alternative to more rigid liners, especially when important settlements are expected. Geomembrane liners provide a quicker installation, an early exploitation of the plant, and in case of accidental damage the possibility to be easily repaired, even underwater, without impacting on the plant operation. The 18 Water Saving Basins of the Third Set of Locks of the Panama Canal expansion project can be considered the first geomembrane application in new pumped storage plants, since they have an average of 5 to 6 fill/empty cycles/day. In this project, several anchorage systems were used to retain the exposed geomembrane liner stable and taut to the surface of the basins under daily varying water levels and against wind uplift. The paper discusses the design, characteristics and advantages of advanced exposed geomembrane systems in recent new projects. At Kokhav Hayarden pumped storage project in Israel, completed in 2022, the anchorage system consists of heat-seaming the geomembrane liner to geomembrane anchor bands embedded in vertical trenches. The connection to the concrete structures (water intakes) consists of a mechanical anchorage designed to accept large settlements and differential displacements. At Abdelmoumen pumped storage project in Morocco, completed in 2023, the concept of heat-seaming the geomembrane liner to geomembrane anchor bands was maintained, while a specific construction method was defined to conform to different embankment materials and subgrade preparation. The liner is a lacquered geomembrane, intended to enhance durability in an environment with particularly high UV radiation. At Pinnapuram pumped storage project in India, currently on-going, the geomembrane liner will be installed on three large embankment dams in the lower reservoir and a 6.6 km long continuous embankment dam in the upper reservoir. The anchorage on the dam face will be obtained by geomembrane anchor bands embedded in vertical trenches and created into a bedding layer of porous concrete.

*Role of Dams and Reservoirs in a Successful Energy Transition – Boes, Droz & Leroy (Eds)*
*© 2023 The Author(s), ISBN 978-1-032-57668-8*

# Ritom HPP – Unforeseen challenges during the inclined shaft excavation

R. Zanoli & S. Massignani
*Marti Tunnel AG, Moosseedorf, Switzerland*

ABSTRACT: The Ritom hydropower plant is located in Switzerland south of the Gotthard massif. Ritom SA, a partner company of the Swiss Federal Railways (75%) and the Canton of Ticino (25%), represented by Azienda Elettrica Ticinese, holds the hydroelectric concession until 2094. The capacity of the over 100-year-old power plant is increased and converted into a pumped storage plant by building a new steel-lined headrace tunnel. The sub-horizontal section (length = 750 m, cross-section = 18.4 m$^2$), excavated in granite by drilling and blasting, encountered a highly fractured section with high water inflows requiring special grouting measures. The following inclined pressure shaft (length = 1.4 km, Outer Diameter = 3.23 m) was excavated by an open hard rock tunnel boring machine specifically configured for this project. The excavation of the shaft with gradients of 42% in the lower section and 90% in the upper section has proven to be one of the biggest challenges of the entire project. This paper focuses on the unforeseen geological challenges encountered during the inclined shaft excavation, and the TBM configurations that allowed them to be safely overcome.

*Theme B: Dams and reservoirs for climate change adaptation*

*Role of Dams and Reservoirs in a Successful Energy Transition – Boes, Droz & Leroy (Eds)*
*© 2023 The Author(s), ISBN 978-1-032-57668-8*

# CRHyME (Climatic Rainfall Hydrogeological Modelling Experiment): A versatile geo-hydrological model for dam siltation evaluation

Andrea Abbate & Leonardo Mancusi
*RSE Ricerca Sistema Energetico, Milano*

ABSTRACT: Dams and reservoirs represent a strategic infrastructure from an energetic viewpoint and their future operativity maintenance is a challenge. Since they interact directly with the surrounding environment, they may encounter siltation problems which undermine the proper functionality. A physical quantification of geo-hydrological processes at the basin scale is a necessary task that hydropower stakeholders require for maintaining the infrastructure functionality. This is particularly true under the projected future climate change scenario where extreme events intensification is expected with high confidence.

The new model concept called CRHyME (Climatic Rainfall Hydrogeological Modelling Experiment) is here presented. This model represents an extended version of the classical spatially distributed rainfall-runoff models. CRHyME model has been written in Python language and it aims to model the effect of geo-hydrological processes occurring at a watershed scale. Knowing the location of a reservoir, the model can quantify the flood and sediment income from the upstream catchment, reconstruct past events or deal with future climate projection.

The CRHyME model, although it is already operational, is continuously updated in order to improve its performance and expand its possible use. Remarkable results have been obtained for the study case of the Valtellina catchment in the Alpine region (northern Lombardy, Italy) where six reservoir siltation ratios have been estimated. CRHyME was also applied considering three different climatic models from the EURO-CORDEX program. The results have highlighted a probable intensification of the geo-hydrological processes across the Alps leading to possible aggravation of reservoir siltation.

# Electronic monitoring of natural hazards prone reservoir regions and catchment areas of Alpine dams

M. Carrel, S. Stähly, J. Gassner & S. Wahlen
*GEOPRAEVENT AG, Zürich*

ABSTRACT: In the Alps, many hydropower infrastructures could potentially be impacted by impulse waves or tsunamis triggered by natural gravitational hazards as landslides, large rock falls or glacier collapse occurring in the vicinity of reservoirs. Impulse waves in reservoirs overtopping dams have the potential to cause serious damage or even complete failure to these key infrastructures. Automatic electronic monitoring of the geomorphological processes involved can provide key information to take measures reducing the impact of natural hazards on these infrastructures. GEOPRAEVENT has been monitoring such hazards in the Alps for more than a decade applying various technologies and different approaches. In this paper, we present different cases of monitoring of slopes, rock cliffs and glaciers that could potentially generate impulse waves in reservoirs located in the Swiss Alps or endanger construction work on hydropower infrastructure. Among others, camera-based systems used to monitor the "Giétro" glacier in the vicinity of the Mauvoisin dam and monitoring of the "Schafselbsanft" area above the Limmeren reservoir with both crack meters installed locally and periodic interferometric radar measurements. These measurement systems provide useful data to monitor the evolution of key processes in the vicinity of these critical infrastructure.

RÉSUMÉ: Dans les Alpes, de nombreuses infrastructures hydroélectriques pourraient être affectées par des ondes d'impulsion ou des tsunamis déclenchés par des risques gravita-tionnels naturels tels que des glissements de terrain, des chutes de pierres importantes ou l'effondrement de glaciers à proximité des réservoirs. Les ondes d'impulsion dans les ré-servoirs qui débordent des barrages peuvent causer de graves dommages, voire une défail-lance complète de ces infrastructures clés. La surveillance électronique automatique des processus géomorphologiques en jeu peut fournir des informations essentielles pour pren-dre des mesures visant à réduire l'impact des risques naturels sur ces infrastructures. GEOPRAEVENT surveille ces risques dans les Alpes depuis plus d'une décennie en utilisant différentes technologies et approches. Dans cet article, nous présentons différents cas de surveillance de pentes, de falaises rocheuses et de glaciers qui pourraient potentiellement générer des vagues d'impulsion dans les réservoirs situés dans les Alpes suisses ou mettre en danger les travaux de construction d'infrastructures hydroélectriques. Entre autres, des systèmes basés sur des caméras ont été utilisés pour surveiller le glacier du "Giétro" à proximité du barrage de Mauvoisin et la surveillance de la zone "Schafselbsanft" audessus du réservoir de Limmeren à l'aide de fissuromètres installés localement et de me-sures radar interférométriques périodiques. Ces systèmes de mesure fournissent des don-nées utiles pour surveiller l'évolution des processus clés à proximité de ces infrastructures critiques.

*Role of Dams and Reservoirs in a Successful Energy Transition – Boes, Droz & Leroy (Eds)*


# Role of water storage reservoirs management and flood mitigation in climate change conditions

T. Dašić
*Faculty of Civil Engineering, University of Belgrade*

ABSTRACT: The impacts of climate change are becoming increasingly pronounced in all aspects of human activity, but are especially evident in the field of water management. One of its most significant consequences is the increasingly pronounced temporal variability of river flows - frequent floods with increasing peak flows and long periods of low water flow. In such conditions, existing flood protection measures are often insufficient to secure the protected area. That is why flood protection systems must be constantly developed, considering their construction, as well as improvement of management measures. The paper presents the consequences of climate change on water resources on the territory of the Serbia. The main principles of water management in such conditions are defined, as well as the role of the estimation of flood hydrographs. The possibilities of applying mathematical models in order to improve the role of active flood protection measures of existing reservoirs are presented. The analyses are performed for water resources systems in the Trebišnjica and Vrbas river basins in the Republic of Srpska (Bosnia and Herzegovina). The main task was to analyse the reduction of peak flow in the urban areas downstream from the analysed reservoirs, taking into account the uncontrolled part of the watershed (between the urban area and the reservoir), from which the torrential tributaries originate. Performed analyses show that reservoirs (even of relatively small active storage) can significantly reduce the peak flow during flood events.

*Role of Dams and Reservoirs in a Successful Energy Transition – Boes, Droz & Leroy (Eds)*


# XFLEX Hydro: Extending operation flexibility at EDF-Hydro Grand Maison PSP

Jean-Louis Drommi, Benoit Joly & Denis Aelbrecht
*Electricité de France, Hydro Engineering Center, France*

Christophe Nicolet & Christian Landry
*Power Vision Engineering, Switzerland*

Cécile Münch & Jean Decaix
*Institute of Sustainable Energy, School of Engineering, HES-SO Valais-Wallis, Sion, Switzerland*

ABSTRACT: Enhanced operation agility is now required from dispatchable generation means, hydro being top of the list. Increasing flexibility of legacy PSP has been implemented at Grand Maison scheme so as to provide regulating power in pump mode thanks to hydraulic short circuit technology. Extensive studies have been achieved to wave operational risk and provide operation boundaries. After 18 months, the demonstration has achieved 2500 hours of operation with full implementation onboard the EDF generating fleet.

*Role of Dams and Reservoirs in a Successful Energy Transition – Boes, Droz & Leroy (Eds)*
*© 2023 The Author(s), ISBN 978-1-032-57668-8*

# Additional water and electricity storage in the Swiss Alps: From studies of potential towards implementation

Jonathan Fauriel
*Head Civil Engineering and Environment, Alpiq Ltd, Switzerland*

Jérôme Filliez
*Senior Civil Engineer, Alpiq Ltd, Switzerland*

David Felix
*Senior Civil Engineer, formerly VAW of ETH Zürich; Aquased GmbH, Switzerland*

Robert M. Boes
*Professor and Head of the Laboratory of Hydraulics, Hydrology and Glaciology (VAW), ETH, Zurich*

ABSTRACT: Storage hydropower from the Swiss Alps contributes considerably to the stability of the European electrical network and is key for the energy transition. Between 2010 and 2020, several research projects were carried out exploring the opportunities for the implementation of new reservoirs in regions of retreating glaciers as well as for the extension of existing reservoirs through dam heightening. The potential for additional electricity storage and production was estimated up to 3.9 TWh and 1.2 TWh/a, respectively.

Within the Energy Strategy 2050 and based on the above-mentioned research, a Round Table, gathering the main stakeholders identified 15 projects to increase the winter electricity production by 2 TWh. These projects have strong connections with existing hydropower schemes, and need to be further developed and refined to meet economic and environmental requirements.

After summarizing the potential studies, the present and future Swiss electricity landscape and the 15 projects of the Round Table, this paper presents a preliminary assessment of their technical risks and unit costs per kWh based on publicly available information. Then, two example projects developed by Alpiq, a main Swiss utility, and related operators are presented, namely the new multi-purpose reservoir at Gorner and the heightening of the Emosson dam.

*Role of Dams and Reservoirs in a Successful Energy Transition – Boes, Droz & Leroy (Eds)*
*© 2023 The Author(s), ISBN 978-1-032-57668-8*

# Guidelines for modelling dam safety adaptation to climate change

J. Fluixa-Sanmartin
*HYDRO Exploitation, Sion, Switzerland*

A. Morales-Torres
*iPresas Risk Analysis S.A., Valencia, Spain*

I. Escuder-Bueno
*Universitat Politècnica de València, Valencia, Spain*
*iPresas Risk Analysis S.A., Valencia, Spain*

ABSTRACT: Risk analysis techniques are useful tools to support decision-making for dam safety by optimizing the economic resources and pointing at the most efficient ways of reducing risks. Climate change is likely to modify dam risks in the future and must be incorporated to long-term safety management strategies to increase economic efficiencies when defining the implementation sequence of risk reduction measures. This article presents a set of guidelines and recommendations to integrate impacts on dam safety, which complexity must depend on the data availability and the depth of the analysis. Three methodologies have been applied to a Spanish dam subjected to the effects of a changing climate, with a focus on hydrological loads. It has been found that simplified methods give valuable information but can oversimplify the reality, while more complex methods provide finer results but at the expense of a higher computational cost. Risk analysis methodologies have proved useful to tackle inherent uncertainties from climate or hydrometeorological information, and to design mitigation measures to counteract the effects of climate change.

*Role of Dams and Reservoirs in a Successful Energy Transition – Boes, Droz & Leroy (Eds)*
*© 2023 The Author(s), ISBN 978-1-032-57668-8*

# Upgrade of the Perlenbach Dam in Germany – A multi-purpose sustainability project

R. Haselsteiner, B. Ersoy & L. Werner
*M&P Water Ltd., Germany*

ABSTRACT: Perlenbach dam is located within the catchment of the Rur River close to the city of Monschau in West Germany at the border to Belgium. The main purpose of the reservoir is water supply, but it also serves the minimum flow requirements, recreation and hydropower.

During the devasting flood incident in July 2021 in western Germany the reservoir faced inflows exceeding a $HQ_{10.000}$ event. Thanks to the conservative design of the existing spillway major harm did not occur. Contrary to the flood event, the 2018 and 2022 dry periods revealed that the capacity of the reservoir is not sufficient to cover such extreme droughts. The water supply association Perlenbach was forced to purchase water from other suppliers of the region.

The existing scheme is a reservoir volume of only 750.000 $m^3$ which represents only 1.7 % of the annual mean inflow. The increase of the reservoir volumes shows positive effects not only onto the water supply security but also the energy production and minimum flow periods. Additionally, the reservoir could contribute to flood retention since the Perlenbach comprises about one third of the Rur catchment at its mouth. Particularly in summer the reservoir could contribute essentially to flood control downstream.

The existing dam is an earth-rockfill dam with an asphaltic surface sealing and a height of 18 m. The upgrade requires the heightening of the existing dam by approx. eight meters. All dam facilities need to be adjusted, the spillway has to be completely reconstructed and existing sediments need to be removed from the reservoir.

Upstream of the reservoir valuable fauna flora habitats are located which are affected by the future reservoir. This impact is a critical aspect and compensation is strongly required. An early participation of all stakeholders is anticipated in order to find a realizable solution.

*Role of Dams and Reservoirs in a Successful Energy Transition – Boes, Droz & Leroy (Eds)*
*© 2023 The Author(s), ISBN 978-1-032-57668-8*

# PKW Spillways: An innovative, resilient and flexible solution from run-off river dams in plains to large dams in mountains suitable for climate change

F. Laugier
*Dam Safety Expert, EDF-CIH, France*

M. Ho Ta Khanh
*Dam Safety Consultant, Vietnam*

J. Vermeulen
*Dam Hydraulics Expert, EDF-CIH, France*

ABSTRACT: Two different examples of Piano Key Weirs spillways (PKW) projects are presented. It highlights the wide range of application and brought benefits: The existing La Raviege dam PKW (40 m high concrete gravity dam, France) initially originally equipped with 2 radial gates and improved with a PKW spillway on crest, and the new Van Phong run-off river dam (Vietnam) equipped with 60 PKW units and 10 large radial gates.

RÉSUMÉ: Le papier présente deux applications très différentes des évacuateurs de crue labyrinthe à touche de piano (PKW). Ces exemples illustrent la très large gamme d'application de ce système et les bénéfices apportées: Le barrage existant de La Raviege (barrage poids de 40 m de haut, France) équipé initialement de 2 vannes radiales et complété par un PKW en crête en rive gauche, et le projet neuf du barrage en rivière de Van Phong (Vietnam) équipé de 60 unités PKW et 10 grandes vannes radiales.

*Role of Dams and Reservoirs in a Successful Energy Transition – Boes, Droz & Leroy (Eds)*


# Enhancing dam safety through contractual stategies

M. Lino
*Dam Expert, ISL Ingénierie, Ciboure, France*

Stéphane Giraud
*Contract Specialist & FIDIC, Adjudicator, Mauguio, France*

Luciano Canale
*Project Director, SCATEC, Naples, Italy*

Bettina Geisseler
*Lawyer, Founder of GEISSELER LAW, Freiburg i.Br., Germany*

ABSTRACT: Dam safety is crucial to successfully ensure the long-term sustainability and the multiple benefits of a dam project and is the primary condition for its acceptance by civil society.

Dam failures worldwide continue to be too numerous, and the International Commission on Large Dams (ICOLD) has recently issued a "World declaration on Dam Safety" [1] and declared dam safety the highest priority of the organization.

The paper looks into dam safety aspects in different phases of the project life cycle and from different stakeholder perspectives, and analyses dam failure records to understand where gaps can be found for the construction of a new dam or the rehabilitation of an existing dam. The statistics on dam failure clearly indicates where corrective measures have to be taken and drives the authors in formulating a set of recommendations, based on their long experience and complementary expertise, to help dam owners, developers and lenders to better address dam safety in their projects. The authors finally analyse the link between dam safety issues and contractual strategy, in order to propose an enhanced contractual and organisational set-up in dam development.

*Role of Dams and Reservoirs in a Successful Energy Transition – Boes, Droz & Leroy (Eds)*
*© 2023 The Author(s), ISBN 978-1-032-57668-8*

# Risk management for the Lago Bianco reservoir in case of a rupture of the Cambrena glacier

J. Maier
*Wettingen, Switzerland*
*previously Swiss Federal Office of Energy, Berne, Switzerland*

R. Baumann
*rebau engineering Ltd., Poschiavo, Switzerland*

H. Stahl
*AFRY Switzerland Ltd., Zurich, Switzerland*

T. Menouillard
*Swiss Federal Office of Energy, Berne, Switzerland*

*Keywords*: Lago Bianco, climate change, glacier rupture, avalanche, impulse flood wave, dam break, reservoir operation

## ABSTRACT

The Lago Bianco reservoir is situated in southeastern Switzerland at an altitude of 2235 m and is used for electricity production. The reservoir with a storage volume of $19 \times 10^6$ m$^3$ is formed by two dams, the 13 m high gravity Arlas dam in the north and the 26 m high arch-gravity Scala dam in the south. The dams were originally built in 1910–12, extended in 1941/42 and renovated in 2000/01.

During an in-depth safety review, which according to Swiss legislation is carried out every fifth year, it was discovered in 2014/15 that the retreating Cambrena glacier had introduced a new risk to the Lago Bianco reservoir: the melting southeastern glacier tongue was slowly moving backwards into a steep rock step and therefore losing its support at the front end and on both sides. According to the glacier specialists there was a danger for a maximum rupture of $1.5 \times 10^6$ m$^3$ office. They estimated that the glacier front will be above the steep rock step around the year 2025 and at that point the possible volume of an ice rupture will decrease down to zero.

Figure 1. Lago Bianco reservoir with the Cambrena glacier, whose southeastern (left) tongue is threatening with a rupture (left photo 2008-08-19, right photo 2022-09-11).

Numerical simulations carried out immediately after the discovery of the new risk showed that a substantial part of the rupturing ice could reach the Lago Bianco reservoir with high speed and initiate an impulse flood wave of up to 4.5 m height directed towards the Arlas dam. For reservoir levels close to the maximum operating level the impulse flood wave could overtop the dam crest and possibly initiate a dam break.

Experiences with other glaciers have shown that an ice rupture is announced over several days or weeks with an increase of the ice flow speed by a factor of 10 or more. A continuous measuring of the ice flow speed allows for a warning and gives time for additional safety measures. Therefore, a photographic camera was installed on the rock above the melting southeastern tongue of the Cambrena glacier. Using the camera pictures, the glacier specialists evaluate the ice flow speed on a daily basis and give immediate warnings, if predefined threshold values are exceeded.

The owner of the Lago Bianco reservoir established an emergency preparedness plan specifically for the case of a rupture of the Cambrena glacier tongue. If the threshold values for the ice flow speed are reached, the reservoir level will be lowered as far down as is necessary to prevent an overtopping of the Arlas dam. For the maximum ice rupture volume of $1.5 \times 106\ m^3$ the safe reservoir level is 2.5 m below the normal maximum operating level. If the ice flow speed increases faster than predicted before the safe reservoir level is reached, the emergency service of the Swiss civil protection will alarm the population in the danger zone and close down both highway and railway passing by the Lago Bianco reservoir.

In 2022, eight years after the discovery of the rupture risk, new photos showed a much smaller Cambrena glacier tongue. Therefore, a new estimate of the maximum ice rupture volume was established. Its result was a rupture volume of less than $0.5 \times 106\ m^3$, for which the impulse flood wave calculation showed a no longer existing danger for the Arlas dam. As a consequence, the restricted safe reservoir level was lifted by the Supervision of Dams Section of the Swiss Federal Office of Energy in November 2022 and the Lago Bianco reservoir is back to a fully normal operation since. The measurements carried out from 2015 until 2022 showed a very good agreement with the prognosis of the glacier specialists.

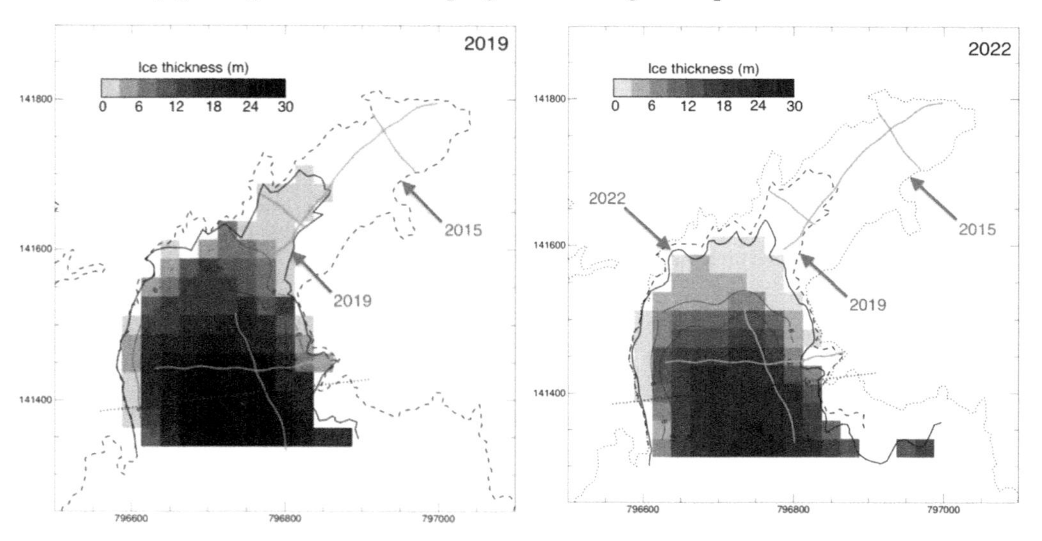


Figure 2. Loss of area and thickness of the Cambrena glacier tongue: situation in 2019 (left) and 2022 (right) (Bauder & Farinotti 2020, 2022).

## REFERENCES

Bauder, A. & Farinotti, D. 2020, unpubl. Eisvolumen der Gletscherzunge des Vadret dal Cambrena. Letter to Repower AG. VAW ETH Zurich.

Bauder, A. & Farinotti, D. 2022, unpubl. Eisvolumen der Gletscherzunge des Vadret dal Cambrena. Letter to Repower AG. VAW ETH Zurich.

*Role of Dams and Reservoirs in a Successful Energy Transition – Boes, Droz & Leroy (Eds)*
*© 2023 The Author(s), ISBN 978-1-032-57668-8*

# Inclinometer monitoring of a dam during hot weather

N. Manzini, S. Van Gorp & Y. Jobard
*SITES SAS, France*

ABSTRACT: The Bazergues Dam is a 156m long and nearly 20m high hydraulic dam located in the Allier department of France. This arch-type dam was built in 1953 and holds a water volume of 1,300 thousand m³ over an area of 14 hectares. Traditionally, the dam's movements have been monitored using direct pendulums. However, these instruments now reach their limit of measurement during very hot and dry conditions, preventing proper monitoring of the structure during these critical periods. To overcome this issue and monitor any movements of the dam, SITES has installed a remote monitoring system using an array of inclinometers. A total of 9 wireless sensors have been deployed on the downstream face of the dam by a team of rope access technicians. These sensors automatically collect data, transmit it to a wireless data logger, which then sends the data to SITES' servers for real-time monitoring with a dedicated interface. This paper focuses on the analysis of data acquired over the past 3 years, including a comparison and correlation of the inclinometers data with ambient conditions, especially during periods of intense heat. A comparison with traditional pendulum measurements is also conducted to validate the monitoring system's data.

RÉSUMÉ: Le barrage de Bazergues est un barrage hydraulique de 156m de long et de presque 20m de hauteur situé dans le département de l'Allier en France. Ce barrage de type voûte a été construit en 1953 et retient un volume d'eau de 1300 milliers de m³ sur une surface de 14 hectares. Les mouvements du barrage sont traditionnellement suivis à l'aide de pendules. Cependant ces derniers atteignent désormais leur limite de domaine de mesure lors des conditions très chaudes et sèches, empêchant le bon suivi de l'ouvrage dans ces périodes critiques. Pour palier cela et suivre d'éventuels mouvements du barrage, SITES a installé et mis en service un système de télésurveillance à l'aide d'une chaîne d'inclinomètres. Au total, une chaîne de 9 capteurs sans fils a été déployée sur le parement en aval par une équipe de cordistes. Les capteurs collectent automatiquement les données, les transmettent vers un collecteur sans-fil, qui transmet les données vers les serveurs de SITES pour un suivi en temps réel avec une interface dédiée. Cet article s'intéresse à l'exploitation des données acquises au cours des 3 dernières années: comparaison et corrélation des inclinomètres avec les conditions ambiantes, en particulier lors des périodes de forte chaleur. Une comparaison avec les mesures de pendule traditionnelle est également réalisée pour valider les mesures du système de monitoring.

*Role of Dams and Reservoirs in a Successful Energy Transition – Boes, Droz & Leroy (Eds)*
*© 2023 The Author(s), ISBN 978-1-032-57668-8*

# Multipurpose dams – A European perspective

A. Palmieri
*Independent Water Infrastructure Adviser, Italy*

D. Maggetti & A. Balestra
*Lombardi Engineering Ltd., Switzerland*

ABSTRACT: The first draft of ICOLD bulletin 171 on Multipurpose Water Storage (MPWS) was circulated in May 2015 during the ICOLD Congress in Stavanger and finally published in 2017. The scope of the bulletin is to provide a view of the dynamics of MPWS in terms of essential elements and emerging trends. In December 2020 a national working group of the Italian Committee on Large Dams (ITCOLD) has been created and started its activities on this topic.

Before the end of the twentieth century, it was predicted (Keller et al. 2000) that one-third of the developing world would have faced severe water shortages by 2025. Since this prediction, it has been observed pressure growing on water resources, with key drivers being more people, growing economies, and increasing hydrological variability.

The true renewable water resource is precipitation, be it in the form of rain or snow. Sporadic, spatial and temporal distribution of precipitation rarely coincides with the demand. Whether the demand is for natural processes or human needs, the only way water supply can match demand is limiting demand itself to match supply, or through the creation of storage to supplement low flow periods.

A ‘Multipurpose’ role is recognized as a key factor in the development of projects involving dams and reservoirs. Sharing benefits between users and needs emphasizes the multiple benefits of dams and reservoirs and fosters acceptance by stakeholders and communities.

For many existing European dams, MPWS has often become a ‘de facto reality’, even for officially single-use dams, because additional requirements are often introduced to the initial single purpose during the life of the works. This introduces constraints that often conflict with each other and with the optimization of the original single purpose. Economic considerations, related to opportunity costs, are also relevant. Therefore, the different uses must be wisely managed, and fairly allocated among different beneficiaries.

A multifunctional approach can better assess the real value of dams and reservoirs, considering both economic and financial aspects.

RÉSUMÉ: La première version du bulletin 171 de la CIGB sur le stockage de l’eau à buts multiples (anglais Multipurpose Water Storage, MPWS) a été diffusée en mai 2015 lors du congrès de la CIGB à Stavanger et a finalement été publiée en 2017. L’objectif du bulletin est de fournir une vue d’ensemble de la dynamique des MPWS en termes d’éléments essentiels et de tendances émergentes. En décembre 2020, un groupe de travail national du Comité italien des grands barrages (ITCOLD) a été créé et a commencé ses activités sur ce sujet.

Avant la fin du vingtième siècle, on prévoyait (Keller et al. 2000) qu’un tiers du monde en développement serait confronté à de graves pénuries d’eau d’ici 2025. Depuis cette prédiction, nous avons observé une pression croissante sur les ressources en eau, dont les principaux facteurs sont l’augmentation de la population, la croissance des économies et le réchauffement climatique.

La véritable ressource en eau renouvelable est la précipitation, que ce soit sous forme de pluie ou de neige. La distribution spatiale et temporelle sporadique des précipitations coïncide rarement avec la demande. Que la demande soit liée à des processus naturels ou à des besoins humains, le seul moyen pour que l’approvisionnement en eau corresponde à la demande est de limiter la demande elle-même pour qu’elle corresponde à l’offre, ou de créer des réservoirs pour compenser les périodes de faible débit.

Le rôle “polyvalent” est reconnu comme un facteur clé dans le développement de projets impliquant des barrages et des réservoirs. Le partage des bénéfices entre les utilisateurs et les besoins met l’accent sur les avantages multiples des barrages et des réservoirs et favorise leur acceptation par les parties prenantes et les communautés.

Pour de nombreux barrages européens existants, le MWPS est souvent devenu une réalité "de facto", même pour les barrages officiellement à usage unique, car des exigences supplémentaires sont souvent ajoutées à l'objectif unique initial au cours de la durée de vie des ouvrages. Cela introduit des contraintes qui entrent souvent en conflit les unes avec les autres et avec l'optimisation de l'objectif unique initial. Des considérations économiques, liées aux coûts d'opportunité, entrent également en ligne de compte. Par conséquent, les différentes exigences doivent être gérées judicieusement et équilibrées entre les différents bénéficiaires.

Une approche multifonctionnelle permet de mieux évaluer la valeur réelle des barrages et des réservoirs, en tenant compte des valeurs économiques et non économiques.

*Role of Dams and Reservoirs in a Successful Energy Transition – Boes, Droz & Leroy (Eds)*
*© 2023 The Author(s), ISBN 978-1-032-57668-8*

# How to evaluate arch dam's behaviour under increased thermal load such as heat waves and extreme cold

E. Robbe & L. Suchier
*EDF-CIH, La Motte-Servolex, France*

A. Simon & T. Guilloteau
*EDF-DTG, Saint-Martin-le-Vinoux, France*

ABSTRACT: Arch dam's behavior is generally strongly influenced by thermal load. In summer, concrete expansion generates displacement of the arch toward upstream, which can lead in some cases to the opening of cracks on the downstream face and instability of gravity abutment. In winter, the arch moves toward downstream, increasing the opening of the dam-foundation interface at the upstream toe for the arch concerned by such issue.

Generally, for the evaluation of the arch dam behaviour during winter or summer load, average seasonal temperatures are used. In 2018, the French guidelines for the safety assessment of existing arch dam behavior proposed to carry out additional analyses which considers increased thermal loads. Such analyses should be performed for dams that show a particular vulnerability to thermal conditions such as very thin arches or arch dams in wide valley which are affected by opening of the dam-foundation interface.

The characterization of increased thermal loads is rather difficult because it is strongly dependent of the inertia of the structure itself: a thin arch dam will react quickly to a short but intense cold weather whereas the behavior of a thicker structure will depend on the temperature of the previous months.

The objective of this communication is to present a uniform methodology based on the monitoring on the dam's crest and statistical analyses involving the HSTT (Hydrostatic Seasonal Temporal Thermal), (Penot,2009) method developed by EDF, to define increased thermal loads with a specified return period. In the same manner that statistical analyses of monitoring data allow to evaluate the radial displacement of the crest of an arch dam during seasonal effect, the methodology allows to define an increase of displacement of the arch under a ten-year return period in winter or summer. Numerical modellings can then be calibrated to represent such an increase of displacement and be used to evaluate the behavior of the dam for such situations.

 *ISBN 978-1-032-57668-8*

# Importance of hydropower reservoirs and dams in Europe to mitigate the energy crisis and to serve as a catalyst and enabler for the Green Deal

A.J. Schleiss
*ICOLD-EPFL, Coordination Team ETIP HYDROPOWER, Switzerland*

J.-J. Fry
*EURCOLD, Coordination Team ETIP HYDROPOWER, France*

M. Morris
*Samui, Coordination Team ETIP HYDROPOWER, France*

ABSTRACT: Hydropower has a long tradition in Europe and contributed significantly during the last century to industrial development and welfare in most of the countries in Europe. The ambitious plan for an energy transition in Europe now seeks to achieve a low-carbon climate-resilient future in a safe and cost-effective way, serving as an example worldwide. The key role of electricity will be strongly reinforced in this energy transition. In many European countries, the phase out of nuclear and coal generation has started with a transition to new renewable sources comprising mainly of solar and wind for electricity generation. However, solar and wind are variable energy sources and difficult to align with demand. The IEA (2021) concludes that hydropower provides "unmatched" flexibility and storage services required for ensuring energy security and delivering more solar and wind power onto the grid. In the future, these services will be in much greater demand to achieve the energy transition in Europe, and worldwide. Hydropower has all the characteristics to serve as an excellent catalyst and enabler for a successful energy transition. The HYDROPOWER EUROPE Forum brought together some 650 relevant stakeholders representing all sectors (including design, construction, production, sectoral associations, environmental and social issues), who participated actively for 40 months (2018-2022) through an extensive program of review and consultation addressing needs of the whole hydropower sector targeting an energy system with high flexibility and renewable energy share. The follow-up ETIP HYDROPOWER project (2022-2025) consolidates the strong network of the HYDROPOWER EUROPE Forum into a sustainable association and helps to unify the voices of hydropower in Europe. ETIP HYDROPOWER will facilitate, enhance and disseminate the Research and Innovation Agenda (RIA) and the Strategic Industry Roadmap (SIR) (drafted under the HYDROPOWER EUROPE project) taking into consideration the future needs of the sector and the R&I targets and the emerging policy priorities. This will help to ensure that hydropower can play the vital role of an enabler in the transition to a clean and safe energy system and the achievement of climate neutrality by mid-century. Furthermore, besides unifying the voices of hydropower, ETIP HYDROPOWER will further align and coordinate the industry RIA and SIR strategies to provide consensus-based strategic advice to the SET Plan (European Strategic Energy Technology Plan) covering analysis of market opportunities and research and development funding needs, biodiversity protection and ecological continuity. Another goal is, in one with Europe and the latest EU climate and energy related policies, deepening the understanding of innovation barriers and the exploitation of research results.

# From energy producer to water manager: A research-industry collaboration

X. Schröder
*Alpiq SA, Lausanne, Switzerland*

E. Reynard
*University of Lausanne, IGD & CIRM, Lausanne, Switzerland*

S. Nahrath
*University of Lausanne, IDHEAP & CIRM, Lausanne, Switzerland*

ABSTRACT: The hydropower industry is expected to play a central role in the double field of energy transition and water security, in a context of climate change, increasing pressure on water resources and geopolitical tensions. The role of alpine hydropower infrastructure in the deployment of a multifunctional water management policy needs to be considered. A research project involving Alpiq and the University of Lausanne is developing this reflection, by clarifying the origin of the concept of multiple use of water and hydraulic infrastructure, the perceptions of multifunctionality by the various water and energy stakeholders, and the issues of governance of resources "water" and "hydraulic infrastructures", particularly in view of the dam relicensing. The work in progress demonstrates the common interest of this collaboration, which is fruitful for both science and industry.

RÉSUMÉ: L'industrie hydroélectrique est amenée à jouer un rôle central dans le double domaine de la transition énergétique et de la sécurité hydrique, dans un contexte de changement climatique, d'augmentation des pressions sur la ressource en eau et de tensions géopolitiques. Des réflexions doivent être menées sur le rôle des infrastructures hydroélectriques alpines dans le développement d'une politique de gestion multifonctionnelle de l'eau. Un projet de recherche associant Alpiq et l'Université de Lausanne développe cette réflexion, en précisant l'origine du concept de multiusage de l'eau et des infrastructures hydrauliques, les perceptions de la multifonctionnalité par les différents acteurs de l'eau et de l'énergie, et les enjeux de gouvernance des ressources « eau » et « infrastructures hydrauliques », notamment dans la perspective des retours de concessions. Les travaux en cours démontrent l'intérêt commun de cette collaboration, fructueuse autant pour la science que pour l'industrie.

*Role of Dams and Reservoirs in a Successful Energy Transition – Boes, Droz & Leroy (Eds)*
 *ISBN 978-1-032-57668-8*

# The use of motor vessels on the reservoirs in Slovenia: A case study on the Sava River

N. Smolar-Žvanut, M. Centa, I. Kavčič & N. Kodre
*Ministry of Natural Resources and Spatial Planning, Slovenian Water Agency, Celje, Slovenia*

ABSTRACT: In accordance with the Slovenian Water Act, the Government of the Republic of Slovenia determines by the Decree individual inland waters or their sections where navigation with the use of motor vessels is permitted, taking into account enabling the general use of water, their protection against pollution and the preservation of the natural balance in the aquatic and riparian ecosystems. Currently, nine Decrees on the use of motor vessels in the river sections are in force in Slovenia. In order to determine the navigation regime in the reservoir of the Brežice Hydropower Plant on the Sava River, a study was made in which the environmental basis from the river basin management plan and bathymetry of the reservoir were taken into account. The indirect and direct impacts of possible pressures due to navigation, on the ecological and chemical status of the water were analyzed. The result of the study is a designated section of reservoir where the navigation with the use of motor vessels is permitted.

*Role of Dams and Reservoirs in a Successful Energy Transition – Boes, Droz & Leroy (Eds)*
*© 2023 The Author(s), ISBN 978-1-032-57668-8*

# Re-operationalization of dams to adapt to climate change in Romania

D. Stematiu
*Romanian Academy of Technical Sciences*

A. Abdulamit
*Technical University of Civil Engineering Bucharest, Romania*

ABSTRACT: Increased hydrologic variability has and will continue to have a profound impact on the water sector through the water availability versus water demand and water allocation. Dam construction is a long-standing strategy to reduce the spatial-temporal variability of natural regime of water. By regulating the water flow, dams alter the natural hydrograph to secure a reliable source of water for a wide variety of human and environmental needs.

The increase in the frequency and intensity of floods and droughts in Romania, combined with the reduced drought and flood storage buffering capacity of dams under a changing climate may have critical inferences for a region's water supply and economy. Building new dams in addition to dams' re-operation may be necessary to balance the climate change impacts on flooding and drought vulnerability.

The present paper deals with several examples of required changes in Romanian dam operation: the new constrains in operation of hydropower developments, the needed increase of flood protection volumes for the existing reservoirs, the new concept in characterizing a flood by its volume instead of the peak inflow and the reasonable procedures to cope with the large siltation of many existing reservoirs.

RÉSUMÉ: La variabilité hydrologique accrue a et continuera d'avoir un impact profond sur le secteur de l'eau à travers la disponibilité par rapport à la demande et allocation de l'eau. La construction de barrages est une stratégie de longue durée visant à réduire la variabilité spatio-temporelle du régime naturel des eaux.

Les augmentations de la fréquence et de l'intensité des inondations et des sécheresses en Roumanie, ainsi que la réduction de la capacité-tampon de stockage des barrages réservoirs en cas de sécheresse et d'inondation dans un climat changeant, peuvent avoir des implications critiques pour l'alimentation en eau et l'économie de la région. Construire des nouveaux barrages et reconsidérer l'exploitation des barrages existants peut être nécessaire pour équilibrer les impacts du changement climatique sur la vulnérabilité aux inondations et à la sécheresse.

L'article présente plusieurs exemples de changements requis dans l'exploitation des barrages roumains: les nouvelles contraintes d'exploitation des aménagements hydroélectriques, l'augmentation nécessaire des volumes de protection contre les crues pour les réservoirs existants, le nouveau concept de caractérisation de l'inondation par son volume au lieu du débit de pointe et les procédures raisonnables pour faire face à la sédimentation importante de certains réservoirs existants.

*Role of Dams and Reservoirs in a Successful Energy Transition – Boes, Droz & Leroy (Eds)*
*© 2023 The Author(s), ISBN 978-1-032-57668-8*

# Design values for dams exceeded: Lessons learnt from the flood event 2021 in Germany

S. Wolf, E. Klopries & H. Schüttrumpf
*Institute of Hydraulic Engineering and Water Resources Management, RWTH Aachen University, Germany*

D. Carstensen
*Institute of Hydraulic Engineering and Water Resources Management, TH Nürnberg, Germany*

R. Gronsfeld
*Waterboard Eifel-Rur (WVER), Düren, Germany*

C. Fischer
*WSW Energy and Water; Wupperverband, Wuppertal, Germany*

ABSTRACT: Large parts of Germany, Belgium and the Netherlands were affected by a severe flood event in mid-July 2021. Dams and reservoirs in the Eifel Mountains reached their design limits and in some cases the design values for a return interval of 1 in 10,000 years were even exceeded. This shows us that planning, design, construction and maintenance of these dams are very safe and that a high level of resilience is guaranteed even in the case of very extreme events.

RÉSUMÉ: Une grande partie de l'Allemagne, de la Belgique et des Pays-Bas a été touchée par de graves inondations à la mi-juillet 2021. Les barrages et les réservoirs des montagnes de l'Eifel ont atteint leurs limites de conception et, dans certains cas, les valeurs de conception pour un intervalle de retour de 1 sur 10 000 ans ont même été dépassées. Cela montre que la planification, la conception, la construction et l'entretien de ces barrages sont très sûrs et qu'un niveau élevé de résilience est garanti même en cas d'événements très extrêmes.

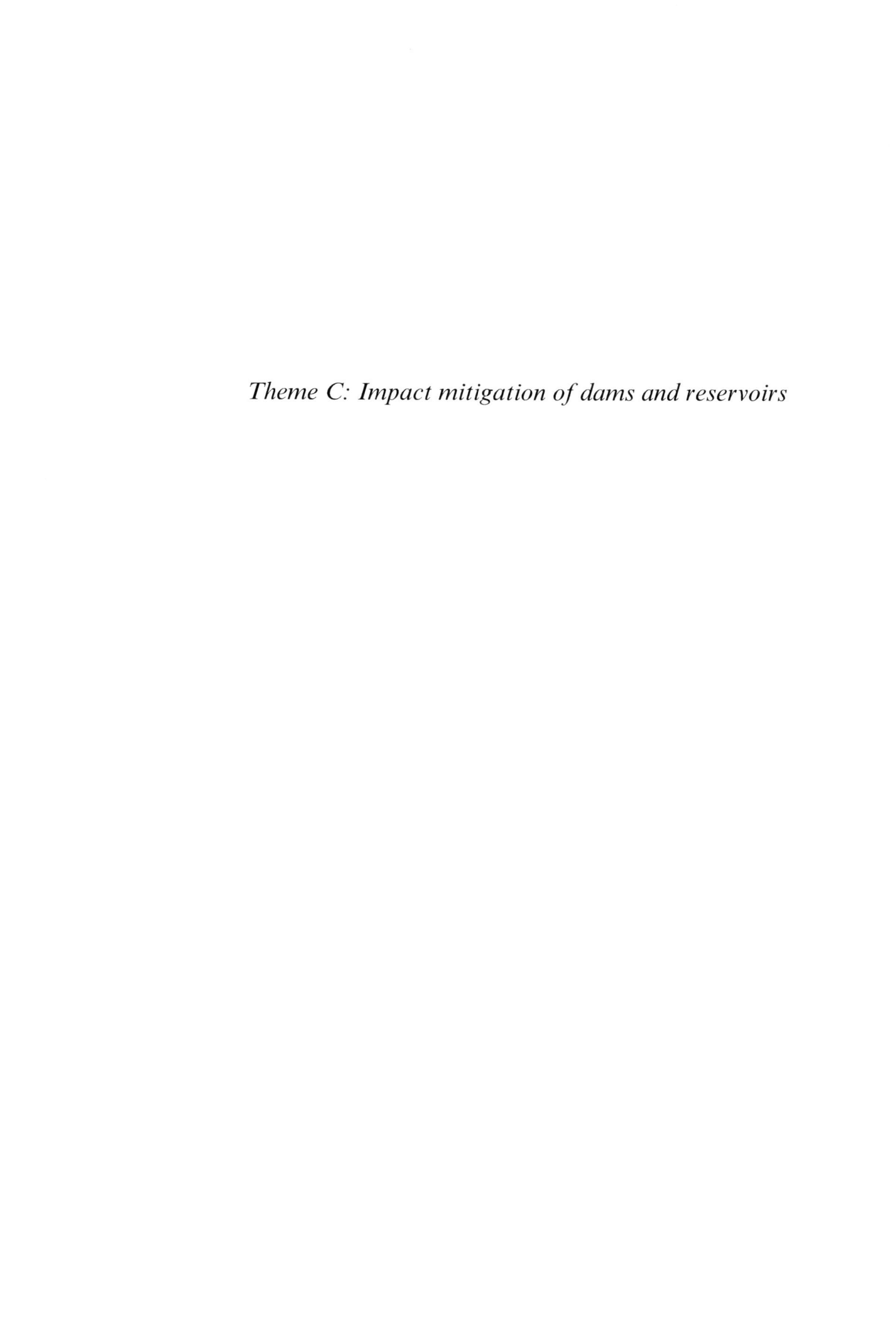

*Theme C: Impact mitigation of dams and reservoirs*

*Role of Dams and Reservoirs in a Successful Energy Transition – Boes, Droz & Leroy (Eds)*


# New Poutès dam (France): Innovative retrofitting to reconcile environment and hydropower

T. Barbier
*Civil Engineer, EDF Hydro Engineering Centre, Le Puy En Velay, France*

S. Lecuna
*Territory Delegate, EDF Hydro, Le Puy En Velay, France*

P. Meunier
*Project Manager, EDF Hydro Engineering Centre, Brive La Gaillarde, France*

ABSTRACT: Poutès dam was at the heart of debates initiated in the 80's for wild river and salmon restoration. Since the expiry of the previous licensing in 2007, Authorities have been looking for a tradeoff solution which might be acceptable to all: EDF Hydro, NGO's and municipalities officials. The objective is to limit the environmental impact as much as possible while maintaining the hydroelectric generation performance.
The renewal of the Monistrol d'Allier hydropower scheme licensing, including Poutès dam, was granted to EDF Hydro in 2015 for a period of 50 years. The new licensing specifications include a significant lowering of the reservoir operating level and the optimizations made to the reconfigured structures ensuring environmental functions.
The reconfiguration of the dam, the work of which took place from 2019 to 2021, consisted of the following tasks:

- Demolishing part of the Creager spillway weirs and piles of the current concrete dam, as well as dismantle the sluices and mechanical components equipping the structure;
- Adapting and lowering the partially demolished dam to get a limited operating level, equipped with a new gated central sluice passage, to ensure free continuity of the structure for sediments and fish during certain periods of the year;
- Reshaping the existing fish pass facilities (for upstream and downstream migration), to adapt them to the new reservoir operating level and improve their performance.

This unusual dam reconfiguration could make it possible to significantly improve fish and sediment transit, and to maintain a performant site's hydroelectric production.

RÉSUMÉ: Le barrage de Poutès est au cœur de discussions engagées dans les années 80 pour le retour à une rivière sauvage sur la thématique du saumon. Depuis l'échéance de la précédente concession en 2007, l'Etat recherche une solution de compromis acceptable par tous: EDF Hydro, les ONG et les élus. L'objectif est de limiter l'impact environnemental tout en maintenant la production hydroélectrique de la chute.

Le nouveau cahier des charges de l'aménagement prévoit un abaissement significatif de la cote de retenue et intègre les optimisations apportées aux ouvrages reconfigurés.

La reconfiguration du barrage, dont les travaux se sont déroulés de 2019 à 2021, a consisté à:

- Démolir une partie des seuils Creager et des piles du barrage actuel, ainsi qu'à démanteler la vantellerie et les organes mécaniques équipant l'ouvrage;
- Aménager le barrage partiellement démolit pour retenir une hauteur d'eau limitée (8 m) et l'agrémenter d'une passe centrale vannée, pour permettre d'assurer la mise en transparence de l'ouvrage (sédimentaire et piscicole) certaines périodes de l'année;
- Reprofiler les ouvrages piscicoles existants (montaison et dévalaison), pour les adapter à la nouvelle cote de retenue et en améliorer le fonctionnement;

Cette reconfiguration atypique permet d'améliorer significativement le transit piscicole et sédimentaire, tout en maintenant la production hydroélectrique du site.

*Role of Dams and Reservoirs in a Successful Energy Transition – Boes, Droz & Leroy (Eds)*
*© 2023 The Author(s), ISBN 978-1-032-57668-8*

# Effects of water releases and sediment supply on a residual flow reach

C. Blanck, R. Schroff & G. De Cesare
*Plateforme de Constructions Hydrauliques (PL-LCH), École Polytechnique Fédérale de Lausanne (EPFL), Lausanne, Switzerland*

ABSTRACT: An artificial floods program has been implemented on the Sarine residual flow reach, downstream of the Rossens dam. Two artificial floods were released in 2016 (coupled with sediment augmentation) and in 2020. A natural flood occurred in 2021. The indicator set on habitat diversity (FOEN) was applied on a Sediment Augmentation and a Control Reach after each flood. Additional data sets comprise the macrohabitats, trout redds and population evolution. The effects of the 2020 artificial flood and the 2021 natural flood on the hydromorphology, the trout reproduction sites and population were compared. To generalize the previous results, hydrological descriptors were analyzed for statistical correlation with the hydromorphological and ecological responses of the river. Results show that the natural flood was larger and had a significantly higher morphological impact, reflected in the increase of the trout redds surface area. The decline in hydromorphological diversity underlines the lack of sediment supply. The correlation analysis confirms the importance of gravel bars as an indicator for trout redds. The role of morphogenic floods for redds availability and hence for trout population is also emphasized. The Sarine residual flow reach's need for regular morphogenic floods, coupled with adequate sediment augmentation measures, is highlighted.

RÉSUMÉ: Un programme de crues artificielles a été implémenté sur le tronçon à débit résiduel de la Sarine, en aval du barrage de Rossens. Deux crues artificielles ont été relâchées en 2016 (couplée à un apport de sédiments) et en 2020. Une crue naturelle a eu lieu en 2021. Le jeu d'indicateurs portant sur la diversité des habitats (OFEV) a été appliqué sur deux tronçons à dépôts de sédiments et de contrôle après chacune des crues. Des jeux de données supplémentaires utilisés sont l'évolution des macrohabitats, les sites de frai et la population des truites. Les effets des crues artificielle (2020) et naturelle (2021) sur l'hydromorphologie, les sites de frai et la population des truites ont été comparés. Pour généraliser ces résultats, des descripteurs hydrologiques ont été analysés pour des corrélations statistiques avec les réponses hydromorphologiques et écologiques de la rivière. Les résultats montrent que la crue naturelle a été plus puissante et a eu un impact morphologique plus important. Cela a été reflété dans l'augmentation de la surface des sites de frai. Le déclin de la diversité hydromorphologique met en évidence le manque d'apport en sédiments. L'analyse de corrélation confirme l'importance des bancs de gravier comme indicateur pour les sites de frai. Le rôle des crues morphogènes pour la disponibilité des sites de frai et donc pour la population de truites est également souligné. Les résultats soulignent le besoin du tronçon à débit résiduel de la Sarine pour des crues morphogènes régulières, combinées à des mesures adéquates d'apports de sédiments.

*Role of Dams and Reservoirs in a Successful Energy Transition – Boes, Droz & Leroy (Eds)*
*© 2023 The Author(s), ISBN 978-1-032-57668-8*

# Bedrock scour prediction downstream of high head dams due to developed rectangular jets plunging into shallow pools

A. Bosman
*Department of Civil Engineering, Stellenbosch University, Stellenbosch, South Africa*

G.R. Basson
*Emeritus Professor, Department of Civil Engineering, Stellenbosch University, Stellenbosch, South Africa*

ABSTRACT: Rock scour formation near the foundation of a dam due to a plunging jet could compromise the safety of the structure. It is therefore essential to predict the geometry of the equilibrium scour hole during the hydraulic design of the dam. Rock scour is normally predicted by analytical-empirical formulae and methods. Despite extensive research since the 1950s, presently there is no universally agreed method to accurately predict the equilibrium scour hole dimensions caused by plunging jets at dams. The main purpose of the research is to contribute to the body of knowledge in predicting the equilibrium geometry of a scour hole in bedrock downstream of a high dam caused by a fully developed rectangular jet plunging into a shallow plunge pool. Both physical and numerical modelling were used to investigate the hydrodynamic and geo-mechanical aspects of rock scour. The physical model investigated the equilibrium scour hole geometries of an open-ended jointed, movable rock bed for different discharges, dam heights, plunge pool depths, rock block sizes, and joint orientations. Novel contributions to science made by the research were the measurement of the dynamic pressures at the joint openings at the water-rock interface of a movable pool bed, while the jet was issued from a rectangular horizontal canal and not a nozzle. From the experimental results, non-dimensional formulae for the scour hole geometry were developed using multi-linear regression analysis. The experimental scour results from this study were compared to various analytical methods found in literature. The equilibrium scour hole depth established in this study best agrees with that predicted by the Critical Pressure method, followed by the Erodibility Index Method.A three-dimensional, multi-phase numerical model, in combination with the developed scour depth regression formula, was used to simulate the equilibrium scour hole geometry in an iterative manner. The proposed three-dimensional numerical model is capable of accurately simulating the scour hole depth, and to a lesser extent the scour length and width.

# Fish passes on the Rhine River – Major structures at EDF Hydro plants to restore fish continuity

Guillian Brousse
*Project Manager, EDF Hydro Engineering Centre, Mulhouse, France*

Régis Thevenet
*Assets Manager, EDF Hydro East Unit, France*

Antoine Vermeille
*Operating Manager, EDF Hydro East Unit, France*

ABSTRACT: At the 16th Rhine river Ministerial Conference on February 13, 2020 in Amsterdam, a new "Rhine 2040" program was adopted. One of the objectives of this program is to complete the restoration of fish continuity between the North Sea and the Schaffhausen falls, a restoration initiated during the Rhine 2020 program with the commissioning of fish passes on Strasbourg and Gerstheim EDF Hydro run-of-river plants.

To do this, two fish passes have been included in the "France Relance Plan" in September 2020 and are in progress: one on Rhinau Hydropower scheme and another one on Marckolsheim Hydropower scheme. These run-of-river hydropower systems each consist of a dam and a diversion canal equipped with shiplocks and a hydro plant. In order to guarantee fish attractiveness, passes are installed downstream of the plant in the tailrace canal and designed to be non-selective with respect to fish species.

The two 'twin' fish passes for Rhinau and Marckolsheim sites, designed by EDF engineering teams, in consultation with local NGO's, the French administration and river foreign stakeholders (Germany, Switzerland), consist of the following structures:

- 8 entrances distributed on the left bank and on the right bank of the tailrace canal, allowing adaptation to river flowrate or level conditions to make the pass as best attractive as possible;
- An attraction flow of 15 m3/s delivered by pumps or turbines depending on the banks;
- A bank-to-bank fish conveying bridge-canal 120 m long allowing fishes to be collected on the right bank;
- A series of basins allowing to pass the plant's head (12 m) and upstream levee up to the headrace canal.

It should be reminded that EDF had initially proposed to study and implement an innovative solution on Rhinau scheme, based on a fish capture-transport and floating barge system, in order to guarantee the deadline for the restoration of fish farming continuity on the Upper Rhine in 2020. This proposal was ultimately not selected by the Authorities.

Fish pass works at Rhinau Hydropower site started in 2022, with a target completion date of 2025. The provisional timetable for Marckolsheim site is 2022-2026.

These major environmental measures and structures will improve fish migration conditions while adapting to the operating conditions of the Hydropower schemes.

*Role of Dams and Reservoirs in a Successful Energy Transition – Boes, Droz & Leroy (Eds)*
*© 2023 The Author(s), ISBN 978-1-032-57668-8*

# Efficiency evaluation and simulation of sediment bypass tunnel operation: Case study solis reservoir

S. Dahal
*Doctoral Researcher, Laboratory of Hydraulics, Hydrology and Glaciology (VAW), ETH Zurich, Switzerland*

M.R. Maddahi
*Visiting Researcher, Laboratory of Hydraulics, Hydrology and Glaciology (VAW), ETH Zurich, Switzerland*
*Doctoral Researcher, Department of Water Engineering, Shahid Bahonar University of Kerman, Iran*

I. Albayrak
*Executive Scientific Collaborator & Lecturer, Laboratory of Hydraulics, Hydrology and Glaciology (VAW), ETH Zurich, Switzerland*

F.M. Evers
*Senior Research Assistant & Lecturer, Laboratory of Hydraulics, Hydrology and Glaciology (VAW), ETH Zurich, Switzerland*

D.F. Vetsch
*Group Head & Lecturer, Laboratory of Hydraulics, Hydrology and Glaciology (VAW), ETH Zurich, Switzerland*

L. Stern
*Operations Manager, Kraftwerke Mittelbünden, ewz, Switzerland*

R.M. Boes
*Director & Professor, Laboratory of Hydraulics, Hydrology and Glaciology (VAW), ETH Zurich, Switzerland*

ABSTRACT: Hydropower is the major source of electricity in Switzerland contributing about 57% (36 TWh/yr) of the total annual generation. Therefore, water storage in hydropower reservoirs is crucial to balance the electricity demand over variable river flow. With the increase in storage demand and climate-related stress it becomes important to sustain the existing reservoir storage capacities. Sedimentation impairs the sustainable operation of reservoirs by reducing the storage volume and may also cause dam safety related issues by the interference of sediment deposits with dam outlets.

Sediment Bypass Tunnels (SBTs) are an effective countermeasure to reduce or even stop sedimentation and contribute to a sustainable use of reservoir storage capacity. This study investigates the performance of an SBT constructed at Solis reservoir in the Swiss Alps, operated by ewz. The SBT was commissioned in 2012 to mitigate continuous propagation of sediment aggradation towards the dam since its construction in 1986. As the inlet of the SBT is located within the reservoir and therefore typically submerged, optimized reservoir operation is required during the intended period of sediment bypassing.

Annual field measurements were conducted to measure the reservoir bathymetry, sediment concentrations, transport rate and sediment particle sizes on the bed to derive the reservoir's sediment balance. The measurements between October 2018 and August 2019 are analyzed to investigate bypass efficiencies of the SBT. The results indicate that the efficiency of the SBT was 80%, and thus considerably higher than the previous efficiency rate of 17%, due to adaptation of the reservoir operation to a lower water level during SBT operation. This implies that with proper synchronization of SBT and reservoir operation, this type of SBT can be highly efficient.

Furthermore, a 1D numerical model is applied to investigate the processes of sedimentation and sediment management for the Solis reservoir. The data from the field measurements is used to set-up, calibrate and validate the model aiming at investigating the performance of the SBT. The model can reproduce the sedimentation as well as SBT operation in terms of longitudinal bed profile evolution and deposition volume. Moreover, the model also allows for simulating additional scenarios, including e.g. no SBT operation, to compare the effects of different operation modes.

*Role of Dams and Reservoirs in a Successful Energy Transition – Boes, Droz & Leroy (Eds)*
*© 2023 The Author(s), ISBN 978-1-032-57668-8*

# Reduction of riverbed clogging related to sediment flushing

R. Dubuis & G. De Cesare
*Platform of Hydraulic Constructions LCH, ENAC, École Polytechnique Fédérale de Lausanne, Lausanne, Switzerland*

ABSTRACT: The ecological effects of dams on sediment and river flow have been subject to an increasing attention, leading to the implementation of mitigation measures such as environmental flow release and sediment replenishment. However, fine sediment dynamics have been subject to less attention. To prevent fine sediment accumulation and maintain reservoir capacity, dam operators conduct sediment flushing operations that can release a significant amount of sediment into the downstream river, potentially damaging the ecosystem and clogging the riverbed. These flushing operations raise questions about their frequency, magnitude, and duration in order to minimize losses to the dam operator while ensuring that the downstream river maintains or improves its ecological services. From a clogging perspective, research is needed to investigate the various options available and their effects on fine sediment dynamics. In this study, the influence of a sediment flushing event on the clogging of riverbeds is analyzed using flume experiments, with a focus on the mobilization of the substrate and the conditions during the falling limb of the hydrograph. Four different scenarios have been tested using silt size sediment in suspension flowing over a bed composed of sand and gravel ranging between 0.1 and 8 mm. Scenarios are characterized by different durations of mobilization phase and falling limb, with different decrease of the suspended sediment concentration during this last phase. The experimental setup allows for the measurement of the permeability of the bed, associated with the presence of an infiltration flow, as well as the vertical distribution of silt in the bed. It appears that the mobilization of the substrate limits the effect of fine sediment on the permeability of the riverbed, although deposition is still taking place. At the end of each flushing event, the permeability was lower in the absence of mobilization. However, mobilization promotes the deposition of fine sediment below the active layer, where fine sediment forms a dense layer. The duration and concentration are key variables to limit the clogging of the substrate below the bed mobilization threshold. When the concentration and flow conditions decrease faster, a smaller reduction of the permeability is observed. Finally, a scenario characterized by low flow conditions followed by the mobilization of the substrate to declog the bed showed a limited impact of the fine sediment deposition while reducing the water volume of the flushing event. This research, although limited to some specific cases, shows that the design of sediment flushing events has an influence on the clogging of the riverbed. The deposition of silt under mobilized bed conditions reveals different results in comparison with a static bed. More research is however needed to take into account the large variety of situations that can arise in riverbeds.

*Role of Dams and Reservoirs in a Successful Energy Transition – Boes, Droz & Leroy (Eds)*
*© 2023 The Author(s), ISBN 978-1-032-57668-8*

# Norwegian sediment handling technologies - recent developments and experiences from projects

Tom Jacobsen
*SediCon AS, Trondheim, Norway*

ABSTRACT: The author has 30 years' experience in designing sediment removal equipment for intakes, desanders, tunnels and reservoirs. Hydrosuction dredging has proven to be an effective way of maintaining reservoir storage, and gravity powered hydrosuction dredges has been supplied tom among others to Tinguiririca and Tricahue reservoirs in Chile in 2022. The capacity of the remotely and automatic operated 400 mm hydrosuction dredge was measured to 183 m3sediments per hour. In 2001 the author designed an ejector dredge with a 160 kW water jetting system and mechanical cutter to disintegrate cohesive clay. The dredge was used to reopen lower parts of the intake at 116 years old Necaxa reservoir in Mexico. Since 2020 the author have developed a boulder excluder which is capable of removing boulders of more than one in size meter from intakes. The boulder excluder is completely without movable parts and still operates autonomously during floods, using only the excess water. A 1,2-meter diameter boulder excluder has been successfully tested at Ulvik power plant in Hardanger in Norway and has since 2020 kept the intake completely clean and free from sediments without any human intervention at all.

# Study on the sedimentation process in Boštanj reservoir, Slovenia

M. Klun, A. Kryžanowski, A. Vidmar & S. Rusjan
*Chair of Hydrology and Hydraulic Engineering, University of Ljubljana, Faculty of Civil and Geodetic Engineering, Ljubljana*

A. Hribar
*HESS, d.o.o., Hidroelektrarne na spodnji Savi, Brežice*

ABSTRACT: The paper presents a study case of Boštanj reservoir on the Sava River in Slovenia. The reservoir for a run-of-the-river dam has been in operation since 2006. The total capacity of the reservoir is 8,000,000 m$^3$, with 1,170,000 m$^3$ of live storage, the maximum operational denivelation in the reservoir is 1 m. The combined rated discharge of the powerhouse is 500 m$^3$/s, while the mean annual discharge of the Sava River is 229 m$^3$/s. The Sava River is a sub-Alpine River with a predominantly torrential character, the discharge conditions in the river can vary substantially, e.g. 3100 m$^3$/s during a 100-year flood. Approximately 3.3 km upstream of the Boštanj dam, in the convex on the right riverbank, a large amount of sediment has been already deposited. During a regular reservoir denivelation, in a length of approximately 400 m, the sediment deposition area is now peaking above the reservoir level. Based on available documentation the material in this part of the reservoir has been already removed in the past, while the aim of this study was to find a permanent solution which would contribute to improved sediment management.

Using the HEC-RAS hydrodynamical model, a full 2D hydraulic model of the reservoir has been constructed. In this study the entire Boštanj reservoir area was considered since local conditions at the convex are affected by global behavior of the sediment in the reservoir. As input data for the hydraulic model a 1×1 m digital terrain model was created using combined lidar and bathymetry measurements. Furthermore, the model was calibrated based on hydrometric measurements in a reference profile. The model consisted of 50,950 calculation cells and the results of the model were used to redesign the area of the reservoir to provide for the sediment continuity and consider the multipurpose use of the reservoir.

Results of the hydraulic model confirm that flow velocities within the area of the reservoir with an increased sediment deposition are low under all flow conditions, meaning the area is generally prone to sediment deposition. Even if the sediment is now completely removed, the natural processes in the river will again start to create a sediment dune in the same location. Therefore, the final solution is designed as a nature-based solution, a plateau of approximately 5900 m$^2$ and 400 m of length, re-designed as a riparian habitat for various aquatic and riparian plants and animals. Instead of fighting the natural processes a redesign of this section of the reservoir is a more appropriate solution; the final solution has been confirmed with a hydraulic model. By raising the plateau level at the mean reservoir level and by providing proper vegetation cover we can create good conditions for the riparian habitat and provide sufficient protection against the erosive power of water.

*Role of Dams and Reservoirs in a Successful Energy Transition – Boes, Droz & Leroy (Eds)*


# Submerged wood detection in a dam reservoir with a narrow multi-beam echo sounder

T. Koshiba
*Kyoto University, Kyoto, Japan*

S. Takata
*Public Works Research Institute, Tsukuba, Japan*

K. Murakami
*Sea Plus Co., Ltd., Yokohama, Japan*

T. Sumi
*Kyoto University, Kyoto, Japan*

ABSTRACT: In 2017, the Susobana Dam in Nagano prefecture became uncontrollable because sediments and submerged driftwood clogged the bottom outlet during flood control. This incident was induced by a combination of large amounts of sedimentation residue which reached the bottom of the gate and driftwood clogged the gate travelling through the flushing water. Such incidents will occur more frequently as now more and more reservoirs are suffering from sedimentation and climate changes, which are a result of these severe flood events nowadays. However, there are few studies on the process of how driftwood is generated, submerged and transported by a flood, thus the driftwood movement during the incident is still a matter of speculation. In this study, as a first step for understanding the dynamics of driftwood movement in reservoirs, we applied the Narrow Multi-Beam Echo Sounder (MBES) system to detect the current distribution of driftwood in the Susobana reservoir.

# HYPOS – Sediment management from space

M. Leite Ribeiro & M. Launay
*GRUNER STUCKY, Switzerland*

K. Schenk & F. von Trentini
*EOMAP, Germany*

M. Bresciani & E. Matta
*Consiglio Nazionale delle Ricerche, Italia*

A. Bartosova & D. Gustafsson
*SMHI, Sweden*

ABSTRACT: The HYPOS tool provides comprehensive, reliable, and scalable access to key hydrological, sediment and water quality parameters to be used in supporting different stages of hydropower installation life cycles. To support the ecological and economical assessment during design and operation phases, HYPOS transforms satellite-based water quality and sediment information together with a worldwide hydrological model, which can be adapted to local conditions at the user's request, and on-site measurements into data useful for decision-making. The water quality and physical parameters supporting the analysis of sediment dynamics in hydropower installations and planning phases are operationally derived from multispectral satellite data from different sensors such as Sentinel-2, Landsat 8/9 or very high-resolution data from Planet SuperDoves. Land use and land cover changes as well as a combination of hydrological and meteorological re-analyses and forecasts complement the analysis. It is showcased in a use case in the Banja and Moglice HPPs, in Albania.

*Role of Dams and Reservoirs in a Successful Energy Transition – Boes, Droz & Leroy (Eds)*
*© 2023 The Author(s), ISBN 978-1-032-57668-8*

# Comprehensive assessment of sediment replenishment and downstream hydro-geomorpho-ecology, case study in the Naka River, Japan and the Buëch River, France

Jiaqi Lin, Sameh A. Kantoush & Tetsuya Sumi
*Water Resources Research Centre, Disaster Prevention Research Institute, Kyoto University, Japan*

ABSTRACT: Sediment replenishment is one common restoration technique to tackle the sediment deficit problem caused by dam construction. The additional sediment supply from replenishment site is beneficial for recovering downstream morphology and aquatic ecosystem. In this paper, we would like to develop a novel methodological approach for assessment of the riverine system during implementation of sediment replenishment in the Naka River, Japan and the Buëch River, France. Key factors for replenishment (placed volume and transported volume, magnitude and duration of flushing flow), and downstream impacts (habitat structures, flow velocity, water depth, riverbed level and substrates, fish species) are considered. Several indices for replenishment and riverine assessment, such as Transported ratio (TR), GUS (Geomorphic Units Survey), BCI (Bed Change Indicator), and Fish Diversity Index (H Value) are determined to statistically evaluate the replenishment works. The results show that the efficiency of SR erosion of France project is higher than Japanese project due to the different optimization strategies. Moreover, both projects can enhance the downstream morpho-ecology by increasing the numbers of GU and promoting the diversity of fish species. It is recommended to conduct continuous SR with adaptable sediment supply to maintain such positive impacts on downstream reach.

*Role of Dams and Reservoirs in a Successful Energy Transition – Boes, Droz & Leroy (Eds)*
*© 2023 The Author(s), ISBN 978-1-032-57668-8*

# Experimental modeling of fine sediment routing: SEDMIX device with thrusters

Montana Marshall, Azin Amini & Giovanni De Cesare
*Platform of Hydraulic Constructions (PL-LCH), EPFL, Switzerland*

ABSTRACT: Reservoir sedimentation is a key challenge for storage sustainability because it causes volume loss, affecting hydropower production capacity, dam safety, and flood management. A preliminary EPFL study proposed and studied an innovative device (called SEDMIX), which uses water jets to keep fine sediments near the dam in suspension and ultimately allows the sediments to be released downstream. The SEDMIX device is composed of two rigid steel parts: one floating and one on the basin bottom holding a multi-jet manifold frame. The jets induce a rotational flow which creates an upward motion and keeps fine sediments in suspension near the dam and water intakes. The sediment can then be continuously released downstream through the power waterways at acceptable concentrations, without additional water loss or required energy. The efficiency of the SEDMIX device has been confirmed through recent experimental simulations and numerical analyses at EPFL. This study involves updated experimental simulations to include thrusters in the device design, instead of jets, because they lead to a less complex arrangement that requires less energy to operate. The experimental setup is similar to the previous experiments (utilizing a glass tank), and tests different thruster parameters to understand the resulting changes in tank turbidity (using turbidity meters).

*Role of Dams and Reservoirs in a Successful Energy Transition – Boes, Droz & Leroy (Eds)*
*© 2023 The Author(s), ISBN 978-1-032-57668-8*

# Dynamic environmental flows using hydrodynamic-based solutions for sustainable hydropower

S. Martel, P. Saharei, G. De Cesare & P. Perona
*Platform of Hydraulic Constructions PL-LCH, ENAC, IIC, EPFL, Switzerland*

ABSTRACT: Water diversions from rivers and torrents for anthropic uses of the resource alter the natural flow regime. As a measure, environmental flows have been prescribed and often are enforced by law to follow policies (e.g., minimal flow, proportional redistribution, etc.) that may vary from geographical location, environmental constraints and country-dependent environmental protection laws. There is not yet general consensus about the optimal release policy to be adopted, although scientific research generally agrees that flow variability for environmental flows should be a hydrological attribute fundamental for maintaining riverine biodiversity. Along with the previous assumption, non-proportional flow release strategies both for small hydropower and for traditional hydropower were proposed to generate optimal solution (*sensu* Pareto) for energy production and riverine ecology. Although non-proportional flow redistribution can easily be achieved by regulating flow devices (e.g., hydraulic gates), there is an interest to avoid the use of blocking structure for both safety control and energy consumption reasons. In this work, we study the redistribution capacity and flow hydrodynamics of asymmetric plate geometries that we propose as a technical and zero-operational-costs solution for generating the desired non-proportional repartition rule. Such plates can be easily mounted to partially cover the metallic rack installed at water intakes and used to intercept the stream while guaranteeing technical and environmental constraints (e.g., minimal and maximal turbined flow, minimal flow release and fish passage scales, transport capacity during high flow, etc.). We perform our study analytically and compare some performances also numerically. In summary, we demonstrate via the proposed analytical framework that different plate geometries correspond to different non-proportional fraction of water left to the environment for varying incoming flow. This work sets the premises for further studies where our approach could be adopted also for design purposes.

*Role of Dams and Reservoirs in a Successful Energy Transition – Boes, Droz & Leroy (Eds)*


# Improving fish protection and downstream movement at the Maigrauge Dam (Switzerland) using an electric barrier

Anita Moldenhauer-Roth
*Laboratory of Hydraulics, Hydrology and Glaciology, ETH Zurich, Switzerland*

Delphine Lambert & Michael Müller
*IUB Engineering Ltd., Berne, Switzerland*

Ismail Albayrak
*Laboratory of Hydraulics, Hydrology and Glaciology, ETH Zurich, Switzerland*

Georges Lauener
*Groupe E Ltd., Granges-Paccot, Switzerland*

ABSTRACT: The main intake of the Maigrauge Dam is protected by a vertically inclined trash rack with 30 mm bar spacing. The bypass entrance for downstream fish passage is located next to the main turbine intake. A PIT tag study showed that many fish approach the bypass but only few enter it. In this project, structural measures were analyzed to improve fish protection at the turbine intake rack and bypass acceptance. To reduce rack passages, two options to electrify the rack were compared: 1) Placing electrodes on the front of the bars of the rack and 2) using the rack as a cathode and placing a row of electrodes 10 cm downstream. Based on numerical simulations of the electric field and comparison to laboratory and literature data, both setups are expected to significantly reduce rack passages. Furthermore, improvements for bypass attraction and acceptance were developed.

*Role of Dams and Reservoirs in a Successful Energy Transition – Boes, Droz & Leroy (Eds)*
*© 2023 The Author(s), ISBN 978-1-032-57668-8*

# Assessment of the hydromorphological effectiveness of sediment augmentation measures downstream of dams

C. Mörtl & G. De Cesare
*Ecole Polytechnique Fédérale de Lausanne, Switzerland*

ABSTRACT: This article presents assessment methods for the hydromorphological effectiveness of sediment augmentation measures downstream of dams. First, we describe different ways of quantifying hydromorphological effectiveness based on typical objectives of sediment augmentation. Then we show how field observations and physical and numerical modelling can be combined to investigate the influence of design criteria and site conditions on hydromorphological evolution following sediment augmentation. We provide examples of the influence of geomorphic units on bedload transport patterns and the influence of augmentation repetition frequency on hydromorphological variability. The results show that geomorphic units can influence deposition patterns and that consecutive sediment augmentation increases morphological and hydromorphological variability. Finally, we discuss the results and summarize some practical implications.

RÉSUMÉ: Notre article présente des méthodes d'évaluation de l'efficacité hydromorphologique des mesures d'augmentation des sédiments en aval des barrages. D'abord, nous décrivons différentes manières de déterminer l'efficacité hydromorphologique en fonction des objectifs typiques de l'augmentation des sédiments. Ensuite, nous montrons comment les observations sur le terrain et la modélisation physique et numérique peuvent être combinées pour étudier l'influence des critères de conception et des conditions du site sur l'évolution hydromorphologique suite à l'augmentation des sédiments. Nous donnons des exemples de l'influence des unités géomorphologiques sur les tendances du transport de charriage et de l'influence de la fréquence de répétition de l'augmentation sur la variabilité hydromorphologique. Les résultats montrent que les unités géomorphologiques peuvent influencer les schémas de déposition et que l'augmentation consécutive des sédiments accroît la variabilité morphologique et hydromorphologique. Enfin, nous discutons les résultats et résumons quelques implications pratiques.

*Role of Dams and Reservoirs in a Successful Energy Transition – Boes, Droz & Leroy (Eds)*
*© 2023 The Author(s), ISBN 978-1-032-57668-8*

# Revitalization of the Salanfe river (Valais, Switzerland): A multi-faced project

M. Perroud & C. Gabbud
*Alpiq, Lausanne, Switzerland*

P. Bianco
*iDEALP, Sion, Switzerland*

J. Rombaldoni
*Bureau d'ingénieur Joël Bochatay Sàrl, St-Maurice, Switzerland*

V. Degen
*Patrick Epiney Ingénieurs Sàrl, Sierre, Switzerland*

ABSTRACT: Due to the construction of the Salanfe SA hydroelectric dam in 1952, the river is disconnected from the upper watershed. On reaching the plain, the river suddenly flows into an artificial bed, between rock banks, an underground sector and a weir at its mouth in the Rhone.

A series of measures resulting from various projects, carried out by different actors, will be implemented over the next 20 years in the Salanfe plain area. The planning and implementation of these measures must be coordinated since the perimeters of these measures and their temporality overlap.

The reflections on the upstream sector of the Salanfe plain area have led to the elaboration of a revitalization project for the watercourse with ambitious fish stakes. The project aims to promote the reproduction and growth of brown trout, and to create a sculpin reserve. The project provides for a widening and lengthening of the watercourse, as well as a diversification of the banks and streambed. Outside the watercourse, a project to safeguard biodiversity, with the creation of ponds favorable to the existing population of yellow-bellied ringers, has been established. A relaxation area, an educational pathway, a waterfall observation mound and paths dedicated to soft mobility are integrated into the overall project.

RÉSUMÉ: De par la construction du barrage de l'aménagement hydroélectrique de Salanfe SA en 1952, la rivière est déconnectée du bassin versant supérieur. En atteignant la plaine, le cours d'eau s'écoule soudainement dans un lit artificiel, entre des berges en enrochements, un secteur souterrain et un seuil à son embouchure dans le Rhône.

Un ensemble de mesures découlant de divers projets, portés par des acteurs différents, verront le jour d'ici à 20 ans dans le périmètre de la Salanfe en plaine. La planification et la mise en œuvre de ces mesures doivent se faire de façon coordonnée, tant les périmètres de celles-ci et leur temporalité se chevauchent.

Les réflexions sur le secteur amont de la Salanfe en plaine ont conduit à l'élaboration d'un projet de revitalisation du cours d'eau aux enjeux piscicoles ambitieux. Il s'agit notamment de favoriser la reproduction et le grossissement de la truite fario et d'aménager une réserve de chabots. Le projet prévoit un élargissement et un allongement du cours d'eau, de même qu'une diversification des berges et du lit. Hors espace cours d'eau, un projet de sauvegarde de la biodiversité, avec la création de mares favorables à la population existante de sonneurs à ventre jaune a été établi. Un espace détente, un sentier didactique, une butte d'observation de la cascade et des chemins dédiés à la mobilité douce sont intégrés au projet global.

*Role of Dams and Reservoirs in a Successful Energy Transition – Boes, Droz & Leroy (Eds)*


# Storage tunnels to mitigate hydropeaking

W. Richter & G. Zenz
*Institute of Hydraulic Engineering and Water Resources Management, Graz University of Technology, Graz, Austria*

ABSTRACT: The optimization of hydraulic systems in diversion tunnel power plants can provide significant economic and ecological benefits. By designing these systems as storage tunnels with differential surge tanks, power plants can be improved to handle surge and sunk compensation to mitigate hydropeaking by optimizing the associated construction costs, allowing for increased flexibility and improved storage management.

*Role of Dams and Reservoirs in a Successful Energy Transition – Boes, Droz & Leroy (Eds)*
*© 2023 The Author(s), ISBN 978-1-032-57668-8*

# Can hydropeaking by small hydropower plants affect fish microhabitat use?

J.M. Santos
*Forest Research Centre, School of Agriculture, University of Lisbon*

R. Leite & M.J. Costa
*CERIS, Instituto Superior Técnico, University of Lisbon*

F.N. Godinho
*Hidroerg, Lisbon, Portugal*

M.M. Portela, A.N. Pinheiro & I. Boavida
*CERIS, Instituto Superior Técnico, University of Lisbon*

ABSTRACT: Despite the numerous benefits of hydropower, this renewable energy source can have serious negative consequences on freshwater fish, as a result of short-term artificial flow fluctuations downstream, often known as hydropeaking. Alterations in species habitat use are expected to occur, as a result of variations in the physical environment. However, such assessments conducted at the microhabitat scale, and stratified by season and ontogeny, have rarely been assessed, yet they are fundamental to improve our mechanistic understanding of hydropeaking influences on fish. The goal of this study was to assess fish microhabitat use and availability of leuciscids at upstream undisturbed (2) and hydropeaking-affected (2) river sites located downstream from two small hydropower instalments (SHP), Douro basin, NE Portugal. Fish surveys (juveniles and adults) were performed in spring and summer by electrofishing followed by the establishment of river transects to acquire use and availability data, respectively. A multivariate approach was then employed to analyse both datasets. Cover and depth were found to be the most important variables driving microhabitat use of species at both the reference and hydropeaking sites. Fish exhibited similar patterns of non-random microhabitat use between the reference and the hydropeaking sites, mainly occupying deeper and more sheltered ones than those available. Overall, seasonal and size-related patterns in species microhabitat use were similar between the reference and hydropeaking sites, with the species showing seasonal patterns in microhabitats use from spring to summer, but, in most of the cases, revealing no size-related difference between both types of sites. This work showed that artificial peak-operations by SHP had negligible effects on fish microhabitat use downstream from SHP when compared to the reference sites, and that the high resilience of the hydropeaking sites appears to be related to the amount of cover habitat and the availability of undisturbed substrates, which provide conditions that still support similar driving patterns of fish habitat use at sites fragmented by SHP.

*Role of Dams and Reservoirs in a Successful Energy Transition – Boes, Droz & Leroy (Eds)*
*© 2023 The Author(s), ISBN 978-1-032-57668-8*

# The Cimia dam in Sicily. A relevant case of rehabilitation

L. Serra
*Waterways, Roma, Italy*

G. Gatto & E. Costantini
*Studio Speri, Rome, Italy*

ABSTRACT: Cimia is an zoned embankment dam with clay core, built in the period 1975-1980 in Sicily. The dam has a maximum height of 39 m and determines a reservoir volume of 10 $Mm^3$. Over the years the structure has undergone a progressive silting due to the influx of solids during floods and actually the bottom discharge is compromised and requires important re-efficiency interventions. The dam shows an excessive lowering of the crest level and cracks due to the consolidation of the embankment. The seismic assessment, carried out accounting for the new Italian guidelines, has shown a potential vulnerability of both the dam and the complimentary works. The planned verification and remediation interventions are complex and expensive, however the work is strategic, as it provides water resources for agriculture. The analysis of these issues and the remedies must therefore involve not only technical verification and planning, but also technical and economic support and training plans that guarantee the sustainability of strategic works. The approach methodology is relevant, given that there are hundreds of dams of the same type, where efforts have been made important for the management of water resources in arid areas, with the aggravated impact of climate change.

*Role of Dams and Reservoirs in a Successful Energy Transition – Boes, Droz & Leroy (Eds)*
*© 2023 The Author(s), ISBN 978-1-032-57668-8*

# Assessing the carbon footprint of pumped storage hydropower – a case study

R.M. Taylor
*Director, RMT Renewables Consulting Ltd, UK*

V. Chanudet
*Environmental Engineer, EDF Hydro Engineering Center, France*

J-L. Drommi
*Electricity Expert, EDF Hydro Engineering Center, France*

D. Aelbrecht
*Head of Technology – Deputy Technical Director, EDF Hydro Engineering Center, France*

ABSTRACT: The role of pumped storage is changing because of the energy transition. Several research initiatives are ongoing to investigate the enhancements in the operational flexibility to support the greater integration of low-carbon variable renewable energies. This paper summarizes a study to estimate the carbon footprint of pumped storage activities. The case of EDF's Grand Maison pumped storage plant was selected. The study presents a methodology for estimating life-cycle emissions which will enable further comparative assessments. These may include the evaluation of the carbon benefit of pumped storage services to the electricity grid, comparison with other pumped storage projects, and with other storage technologies.

RÉSUMÉ: La transition énergétique amène un nouveau regard vis-à-vis du rôle des station de transfert d'énergie par pompage (STEP). De nombreuses études sont en cours pour améliorer leur flexibilité d'exploitation afin de permette une meilleure intégration des énergie renouvelables faiblement carbonées. Cet article décrit une étude visant à estimer l'empreinte carbone des STEP avec comme cas d'usage l'aménagement de Grand Maison exploité par EDF. Cette étude présente une méthodologie d'estimation des émissions sur le cycle de vie de l'aménagement, base pour des analyses comparatives ultérieures. Ces dernières peuvent inclure l'évaluation du gain en carbone apporté par les services d'une STEP sur le système électrique, la comparaison avec d'autres projets de STEP ou avec d'autres technologies de stockage d'énergie.

*Role of Dams and Reservoirs in a Successful Energy Transition – Boes, Droz & Leroy (Eds)*


# Exploring the efficacy of reservoir fine sediment management measures through numerical simulations

S. Vorlet, M. Marshall, A. Amini & G. De Cesare
*Hydraulic Constructions Platform, Ecole Polytechnique Fédérale de Lausanne (EPFL), Lausanne, Switzerland*

ABSTRACT: Reservoir sedimentation is one of the main challenges in the sustainable operation of large reservoirs because it causes volume loss, affecting hydropower production capacity, dam safety, and flood management. To ensure the sustainability of deep reservoirs by maintaining sediment flow continuity, it is essential to understand the mechanisms of the sedimentation process. The prediction of sediment deposition can enable adequate sediment management, including the design and implementation of prevention and mitigation measures. Different countermeasures are now being used to tackle sedimentation problems. However, many of these measures have a considerable ecologic and/or economic impact. It is therefore paramount to develop new efficient measures to ensure fine sediment transport through large reservoirs. This paper presents three innovative measures for fine sediment management in reservoirs and summarizes how state-of-the-art numerical modeling might help to assess the efficiency of these measures.

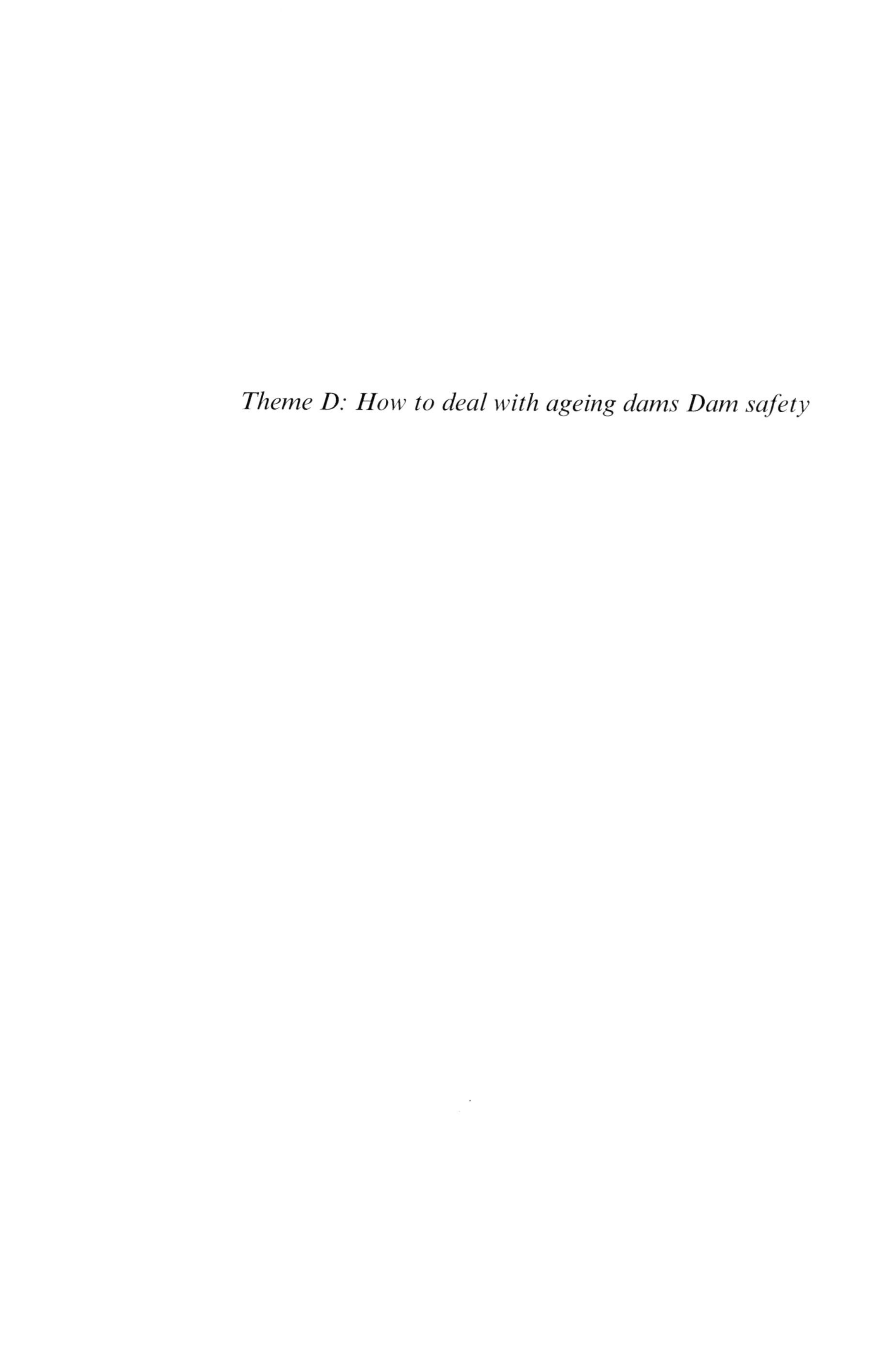

*Theme D: How to deal with ageing dams Dam safety*

*Role of Dams and Reservoirs in a Successful Energy Transition – Boes, Droz & Leroy (Eds)*
*© 2023 The Author(s), ISBN 978-1-032-57668-8*

# Ageing dams in Switzerland: Feedbacks of several case studies

Nicolas J. Adam
*Civil Engineer, Alpiq Ltd, Switzerland*

Jérôme Filliez
*Senior Civil Engineer, Alpiq Ltd, Switzerland*

Jonathan Fauriel
*Head Civil Engineering and Environment, Alpiq Ltd, Switzerland*

ABSTRACT: Swiss dams, mainly built for electricity generation, are monitored by an organization composed by the dam owner, the dam operator, qualified professionals, and the Swiss Federal Office of Energy.

The paper presents different case studies of how ageing dams is addressed, including dams affected by alkali-silica reaction (ASR) or subjected to a change in thermal loading. The panel of case studies allows to identify the different stages of ageing and highlight the investigation process. It includes the deviation of the structural behavior, the crack appearance, and the detailed on-site investigation. These feedbacks challenge the existing processes, the studies used to help the decision and highlight strengths and weaknesses and when it is necessary to act. Furthermore, different solutions are highlighted from the rebuild or reinforcement to a reinforced monitoring.

The synthesis suggests a rehabilitation decision process to ensure an adequate response for both safety and economical aspects. The rehabilitation monitoring is finally discussed to control the effectiveness of the owner's response.

*Role of Dams and Reservoirs in a Successful Energy Transition – Boes, Droz & Leroy (Eds)*
 *ISBN 978-1-032-57668-8*

# Geological hazard evaluation for the dams constructed at Drin valley

S. Allkja, A. Malaj & K. Petriti
*A.L.T.E.A. & Geostudio 2000, Tirana, Albania*

ABSTRACT: The Drin River valley is very important for Albania, on the energy sector, and electricity production. The most important dams constructed in this valley are Fierza, Komani, Qyrsaqi, Rragami, Zadeja dam and the newest one (currently under study process), Skavica dam. This article will present some of our findings regarding the geological hazards identified on these dams, give recommendations to reduce the risk and keep them under surveillance.

RÉSUMÉ: Le Drin est un élément clé de l'économie Albanaise, notamment pour sa production électrique. Relativement, les plus barrages importants ont été construits sur cette vallée. Nous pouvons mentionner, Fierza, Komani, Qyrsaqi, Rragami, le barrage de Zadeja et aussi le plus récent (actuellement à l'étude), le barrage de Skavica. Nous allons traiter des phénomènes géologiques et des risques identifiés dans les zones où ces barrages sont construits et nous donnerons des recommandations pour réduire le risque et maintenir sous surveillance et exploitation plus longtemps.

*Role of Dams and Reservoirs in a Successful Energy Transition – Boes, Droz & Leroy (Eds)*
*© 2023 The Author(s), ISBN 978-1-032-57668-8*

# Galens arch dam strengthening works

F. Andrian, N. Ulrich & P. Agresti
*ARTELIA, Grenoble, France*

Y. Fournié
*SHEM, Toulouse, France*

ABSTRACT: The Galens dam is a 20m high thin arch dam built in a wide valley. Because of this unfavorable configuration, a transverse crack has developed at the base of the central cantilevers near the contact with the ground level. Moreover, the higher-than-usual monitored displacements caught the attention of the dam operator, the Société Hydro-Électrique du Midi (SHEM). These displacements were further amplified by the thermal sensitivity of the arch, especially during winter in one of the coldest regions of France. The stability analyses conducted by ARTELIA from 2014 revealed a non-conformity during rare winter loadings. Although the short and mid-term safety was not compromised, the strengthening works of the dam was considered by the SHEM. In the first part, the paper will describe the studies that led to the choice of the strengthening principle. The "cyclic calculation" method will be discussed, an approach which makes it possible to model one aspect related to the ageing of the dam. The reasons that led to the choice of the strengthening solution will also be discussed. In the second part, the technical specificities involved in the works will be described. The mass concrete mix optimization, the bonding between the new mass concrete to the existing arch, the simplified groutable contraction joints and the upgrade of the monitoring system will be detailed.

RÉSUMÉ: Le barrage des Galens est une voûte mince en vallée large d'une vingtaine de mètres de hauteur. A cause de cette configuration défavorable, une fissure traversante s'est développée en pied des consoles centrales, près du contact avec le terrain naturel. De plus, les déplacements auscultés inhabituellement élevés ont attiré l'attention de l'exploitant, la Société Hydro-Électrique du Midi. Ces déplacements sont par ailleurs amplifiés par l'effet thermique, notamment pendant l'hiver dans l'une des régions les plus froide de France. L'étude de stabilité menée par ARTELIA dès 2014 a montré une non-conformité pour le chargement hivernal rare. Bien que la sécurité à court et moyen terme ne fût pas compromise, des travaux de renforcement ont alors été envisagés par la SHEM. Dans la première partie, les études qui ont menées au choix du principe de confortement seront décrites. La méthode de « calculs cycliques », permettant de simuler le vieillissement de la voûte, sera abordée. Le choix de la solution de confortement sera également expliqué au travers des avantages et inconvénients de chaque option.

Dans la seconde partie, les spécificités techniques liées aux travaux seront abordées. En particulier, le choix de la formulation du béton, la jonction avec la voûte existante, les joints de dilation injectables, et l'amélioration du dispositif d'auscultation seront les points discutés.

# Design concept for sustainable cut-off walls made of highly deformable filling materials

K. Beckhaus
*BAUER Spezialtiefbau GmbH, Germany*

J. Kayser
*Federal Waterways Engineering and Research Institute, Germany*

F. Kleist
*SKI GmbH + Co.KG, Germany*

J. Quarg-Vonscheidt
*Koblenz University of Applied Sciences, Germany*

D. Alós Shepherd
*Karlsruhe Institute of Technology, Germany*

ABSTRACT: Currently, a joint working group of the German Association for Water, Wastewater and Waste (DWA), the Hafentechnische Gesellschaft e.V. (HTG) and the Deutsche Gesellschaft für Geotechnik e.V. (DGGT) is working on recommendations for a new design concept for sustainable cut-off walls made of "highly deformable cut-off wall materials". The focus lays essentially on a higher utilisation of the visco-elastic and also plastic deformation capacity of low-strength cement-bound filling materials. This allows for a lower required cement content and thus offers a sustainable advantage, as the equivalent carbon footprint will be significantly lower. Compared to the classical design method, a lower compressive strength shall be allowed. This should be high enough to ensure erosion safety, but with the lowest possible deformation modulus, while at the same time allowing deformation without exceeding the permissible compressive and tensile stresses. The article presents how this high deformation capacity can be applied in the design and thus ensure that more economical and resource-saving cut-off walls can be planned.

RÉSUMÉ: Actuellement, un groupe de travail conjoint d'associations allemandes concernées (DWA, HTG et DGGT) travaille à l'élaboration de recommandations pour un nouveau concept de design de parois moulées durables faites de "matériaux de parois moulées hautement déformables". L'accent est mis essentiellement sur une meilleure utilisation de la capacité de déformation viscoélastique et plastique des matériaux cimentaires à faible résistance. Cela permet de réduire la teneur en ciment nécessaire et offre donc un avantage environnemental, puisque l'empreinte CO2 équivalente sera considérablement réduite. Par rapport à la méthode de conception classique, une résistance à la compression plus faible peut être autorisée. Cette résistance à la compression doit être suffisamment élevée pour garantir la sécurité contre l'érosion, mais avec le module de déformation le plus bas possible, tout en permettant une déformation sans dépasser les contraintes de compression et de traction admissibles. L'article présente la manière dont cette capacité de déformation élevée peut être appliquée lors de la conception, ce qui permet de planifier des murs de soutènement plus économiques et plus économes en ressources.

*Role of Dams and Reservoirs in a Successful Energy Transition – Boes, Droz & Leroy (Eds)*
*© 2023 The Author(s), ISBN 978-1-032-57668-8*

# Digitalization for a targeted and efficient dam management

F. Besseghini, C. Gianora, M. Katterbach & R. Stucchi
*Lombardi Engineering Ltd., Giubiasco, Ticino, Switzerland*

ABSTRACT: Dams are not only extraordinary engineering structures; they play a crucial role in the management of our most important vital resource, namely water. Currently, the expression "energy crisis" is familiar even to non-professionals. This is an important issue that plant operators are confronted with. A shutdown of the plant or lowering of the reservoir for extraordinary maintenance or to mitigate water scarcity, leads to major economic losses, both for dam owners, and homeowners, who are facing rising electricity costs. Is there a better way to adequately manage dams and appurtenant structures? The key lies in digitalization. Digitalization becomes more and more important to operate and maintain the facilities efficiently and adequately. The implementation of digital models allows the plant owner to undertake appropriate and faster business decisions and, therefore, to proactively manage the risks associated with aging dams' infrastructure without affecting their operation. Digital models allow to overcome a common problem in dam monitoring: the dispersion of a large part of data volumes in multiple archives. In this way, all monitoring data can be kept in a single platform and effectively and efficiently visualized. These models also make real-time monitoring possible, thanks to the implementation of remote sensors for monitoring. New technologies, along with deterministic predictive models, enrich the digital twin and create a more complete picture over time without impeding the plant operation. It enables proper life cycle management of the structure, resulting in an exponential rise in the model's efficacy over time.

*Role of Dams and Reservoirs in a Successful Energy Transition – Boes, Droz & Leroy (Eds)*
*© 2023 The Author(s), ISBN 978-1-032-57668-8*

# Digital cloud-based platform to predict rock scour at high-head dams

E.F.R. Bollaert
*AquaVision Engineering Llc, Ecublens, Switzerland*

ABSTRACT: Rock scour downstream of high-head dams and in unlined channels and stilling basins of dam spillways is a more and more frequently occurring phenomenon. Climate change and related regulatory requirements generate frequent functioning of dam spillways, and sometimes triggers a first main spillage of spillways that have been constructed decades ago but were never used, or only used for minor discharges.

Hence, unlined rock masses downstream of those structures experience more and more hydraulic stress by action of hydrodynamic pressures at the water-rock interface. Sound prediction of rock scour potential becomes more and more pertinent, especially for cases where the scour hole may potentially regress towards the dam or spillway foundations.

This paper first presents novel computational methods for the prediction of scour of unprotected rock in plunge pools downstream of high-head dams. These novel methods are based on flow velocities, applicable stream power or dynamic pressures at the water-rock interface and complete the existing Comprehensive Scour Model originally developed by Bollaert (2002) and Bollaert & Schleiss (2005).

The computational methods have been bundled into a cloud-based numerical platform, rocsc@r, based on user-defined tailored parametric settings of the dam and spillway geometry, the turbulent flow and the rock mass. This platform allows one to determine scour formation and plunge pool shape in a quasi-3D manner, as a function of intensity and time duration of overflows. Moreover, the scour methods can be directly coupled with the FLOW-3D® CFD software for a sequential fluid-solid coupling.

Second, the paper presents the application of these novel computational methods to the case of a significant scour hole that was observed in the unlined rock mass of the plunge pool downstream of Chucàs Dam in Costa Rica, following the 2017 tropical storm flood event with a return period of about 100 years.

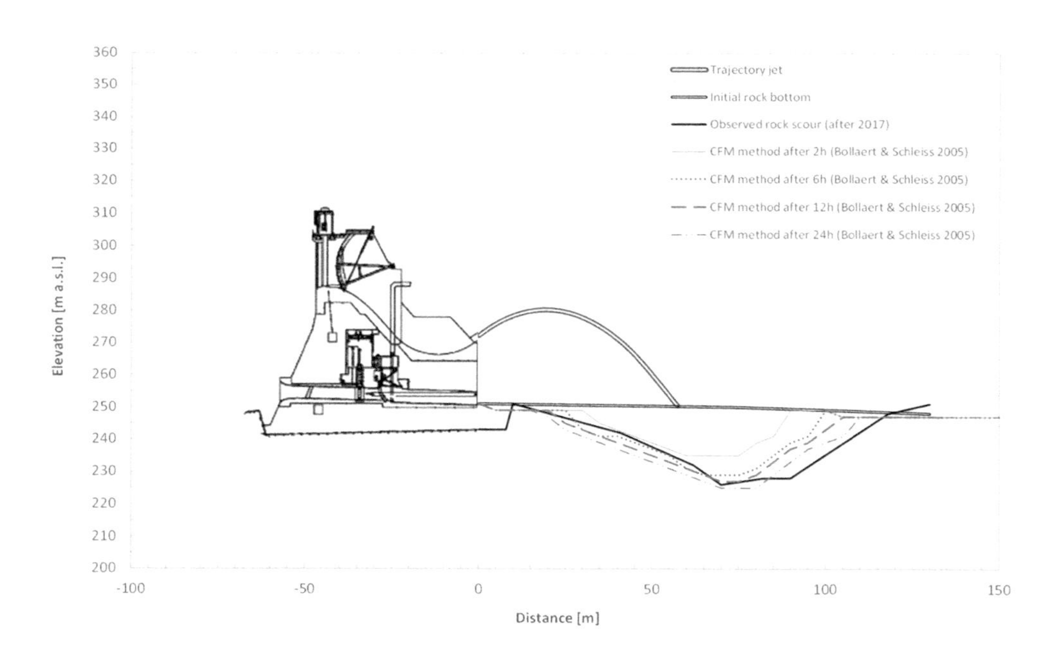


Figure 1. UP: Jet trajectory of Chucàs Dam during 2017 flood (source: O. Jiménez); DOWN: Scour potential computed by rocsc@r software and comparison with in-situ measured scour.

*Role of Dams and Reservoirs in a Successful Energy Transition – Boes, Droz & Leroy (Eds)*
*© 2023 The Author(s), ISBN 978-1-032-57668-8*

# Sharing elements of EDF feedback on the operation and maintenance of pendulums

P. Bourgey, T. Guilloteau & J. Sausse
*Dam safety expert, EDF, France*

ABSTRACT: Although pendulums are widely used in monitoring for their reliability, precision and robustness, the fact remains that they can present material and functional problems which require regular control and maintenance actions, or even replacement in some cases. Through concrete and illustrated cases, this article presents in a qualitative way the different problems encountered.

RÉSUMÉ: Si les pendules sont des appareils très utilisés en auscultation pour leur fiabilité, leur précision et leur robustesse, il n'en demeure pas moins qu'ils peuvent présenter des problèmes matériels et fonctionnels, qui nécessitent des actions régulières de contrôle et de maintenance, voire de remplacement dans certains cas. Au travers de cas concrets et illustrés, cet article présente de manière qualitative les différentes problématiques rencontrées.

*Role of Dams and Reservoirs in a Successful Energy Transition – Boes, Droz & Leroy (Eds)*
*© 2023 The Author(s), ISBN 978-1-032-57668-8*

# Lifetime analysis of the Sta. Maria arch dam behaviour

M. Bühlmann, S. Malla & R. Senti
*Axpo Power AG, Baden, Switzerland*

ABSTRACT: The Sta. Maria arch dam was built between 1965 and 1967. After 40 years of operation, the Gotthard Base Tunnel (GBT) was constructed in the area of the dam. The tunnelling caused rock mass displacements, leading to irreversible deformations of the dam. To study the current behaviour of the dam as well as the effects of the tunnelling, a statistical analysis of the dam deformations was performed. For each of the 28 measurement stations, a statistical model was set up and calibrated. To analyse the entire lifetime of the dam considering the concrete temperature data, which are available only since 1980, a novel approach was applied: the temperature effects were simulated by a seasonal function until 1980 and later the measured temperatures were considered. After calibration, the contributions of water load, temperature and time-dependent effects could be separated. These results were used to determine the adjusted behaviour indicators (ABI), which can be employed as a powerful tool to evaluate dam behaviour over the whole lifetime. The case study presented in this paper shows that tunnelling is responsible for around 25 % of the irreversible dam deformations since the first impoundment.

RÉSUMÉ: Le barrage-voûte de Sta. Maria a été construit entre 1965 et 1967. Après 40 ans d'exploitation, le tunnel de base du Saint-Gothard (GBT) a été construit dans la zone du barrage. Le creusement du tunnel a provoqué des déplacements de la masse rocheuse, entraînant des déformations irréversibles du barrage. Afin d'étudier le comportement actuel du barrage ainsi que les effets du creusement du tunnel, une analyse statistique des déformations du barrage a été réalisée. Pour chacune des 28 stations de mesure, un modèle statistique a été mis en place et calibré. Pour analyser la durée de vie du barrage en tenant compte des données de température du béton, qui ne sont disponibles que depuis 1980, une nouvelle approche a été appliquée: les effets de la température ont été simulés par une fonction saisonnière jusqu'en 1980 et, par la suite, les températures mesurées ont été prises en compte. Après le calibrage, les contributions de la charge d'eau, de la température et des effets dépendant du temps ont pu être séparées. Ces résultats ont été utilisés pour déterminer les indicateurs de comportement ajustés (ABI), qui peuvent être utilisés comme un outil puissant pour évaluer le comportement du barrage pendant toute sa durée de vie. L'étude de cas présentée dans cet article montre que le creusement de tunnels est responsable d'environ 25 % des déformations irréversibles du barrage depuis la première mise en eau.

Traduit avec www.DeepL.com/Translator (version gratuite).

# Effect of invert roughness on smooth spillway chute flow

M. Bürgler, D.F. Vetsch, R.M. Boes & B. Hohermuth
*Laboratory of Hydraulics, Hydrology and Glaciology (VAW), ETH Zurich, Zurich, Switzerland*

D. Valero
*Institute for Water and River Basin Development (IWG), Karlsruhe Institute of Technology (KIT), Karlsruhe, Germany*
*Water Resources and Ecosystems department, IHE Delft, Delft, The Netherlands*

ABSTRACT: Spillway chutes are appurtenant dam outlet structures with the purpose to safely convey large discharges during extreme flood events. During such events, hydraulics plays a major role in the safety of the structure. Along a spillway chute, water is accelerated by gravity and may reach flow velocities in the order of 10 to 50 m/s, implying a considerable cavitation risk. On the spillway invert, turbulence is generated by shear stresses and surface roughness, which results in self-aeration of the flow once the turbulent boundary layer interacts with the free surface. For reliable design guidelines of spillways, knowledge of air concentrations along the spillway chute is essential, as entrained air concentrations can mitigate the risk of cavitation at the expense of risking overtopping of the chute walls due to flow bulking, or further accelerating the flow due to drag reduction. While it is well known that the invert roughness is the controlling parameter for boundary layer development and the self-aeration process (for a given slope and discharge), the quantitative understanding of roughness effects on air-water flow properties is still limited by the availability of data sets that target this variable. In this research, the effects of invert roughness on smooth spillway chute flow are investigated in a large-scale physical model. The investigated flow properties include the clear water and air-water mixture flow depths, depth-averaged flow velocities, air concentrations, and friction factors. Based on the experimental data, we demonstrate that the streamwise development of depth-averaged air concentration is significantly affected by invert roughness, which in turn also affects the bottom air concentration downstream of the inception point. Further, we found that friction factors are significantly affected by the relative boundary layer thickness in the developing non-aerated flow region, but also by bottom air concentrations in the aerated flow region. Good agreement between experimentally determined friction factors and established theoretical relations was found. Overall, our findings contribute to a qualitative description of invert roughness effects on air-water flow properties for a robust design of spillways, thus contributing to safer dam infrastructure.

*Role of Dams and Reservoirs in a Successful Energy Transition – Boes, Droz & Leroy (Eds)*
*© 2023 The Author(s), ISBN 978-1-032-57668-8*

# Nonlinear deterministic model for a double-curvature arch dam

E. Catalano & R. Stucchi
*Lombardi Engineering Ltd., Switzerland*

ABSTRACT: Deterministic models allow predicting dam response during the structure lifecycle. They are usually based on structural models and assume linear elastic behavior for the structure. However, model precision and reliability could be affected whenever nonlinear effects influence the structure response. We present the case of Emosson dam where nonlinear behaviour of vertical construction joints was found to influence the dam response. Including such aspect in the model enhanced its precision but limited its applicability.

RÉSUMÉ: Les modèles déterministes permettent de prédire la réponse du barrage pendant la durée de vie de la structure. Ils sont généralement basés sur des modèles structuraux et supposent un comportement élastique linéaire pour la structure. Une telle hypothèse pourrait affecter la précision et la fiabilité du modèle lorsque des effets non linéaires influencent la réponse de la structure. Nous présentons le cas du barrage d'Emosson où le comportement non linéaire des joints de construction verticaux s'est avéré influencer la réponse du barrage. L'inclusion de cet aspect dans le modèle a amélioré sa précision tout en limitant son applicabilité.

*Role of Dams and Reservoirs in a Successful Energy Transition – Boes, Droz & Leroy (Eds)*
*© 2023 The Author(s), ISBN 978-1-032-57668-8*

# Breach analysis of the Lozorno II. Dam

E. Cheresova, T. Mészáros & M. Mrva
*VODOHOSPODÁRSKA VÝSTAVBA, SOE, Bratislava, Slovakia*

ABSTRACT: The main objective of the reported manuscript is to analyze the breach of an earthfill embankment, specifically the Lozorno II. dam, serving for further purposes to possible reclassification of the hazard category of the dam. The Lozorno II. dam is located in the Slovak repub-lic, lying in the Záhorská Lowland, where the need for reclassification results from intensive urban development downstream of the dam. For our work purposes we developed an HEC-RAS model from a digital terrain model (DTM) creating geospatial data in RAS Mapper to represent the river system and floodplain area specifically for performing a dam breach analysis with HEC-RAS Version 6.3.1. Results of the analysis, and the brief discussion on it are included in the presented paper.

*Role of Dams and Reservoirs in a Successful Energy Transition – Boes, Droz & Leroy (Eds)*
*© 2023 The Author(s), ISBN 978-1-032-57668-8*

# Dam safety and surveillance: Return of experience from the perspective of the Swiss Federal authorities

M. Côté, R.M. Gunn & T. Menouillard
*Swiss Federal Office of Energy, Dam Safety and Surveillance Section, Bern, Switzerland*

ABSTRACT: This paper provides a return of experience from the perspective of the Dam Section of the Swiss Federal Office of Energy, SFOE, on a range of challenging safety and surveillance issues that the Swiss dam industry is facing today. These include the implementation of new regulations, pumped storage schemes, flood verifications, dam ageing, climate change, the prioritization of safety measures, reservoir slope stability, sedimentation, hydraulic discharge structures, bi-national safety and surveillance requirements, the impact of solar photovoltaic panels on dam safety, surveillance and other subjects. In accordance with Art. 29 of the WRFO (Water Retaining Facilities Ordinance, Status on the 1st January 2023) the main objectives of the paper are to ensure the exchange of information at an international level, secure expertise, promote research and to outline some of the tasks undertaken by the Swiss authorities, putting the emphasis on challenging problems. The conclusions allow dam owners and operators to perhaps reassess their dam surveillance and safety concepts and take appropriate measures in a timely manner.

*Role of Dams and Reservoirs in a Successful Energy Transition – Boes, Droz & Leroy (Eds)*


# Failures and incidents in Greek dams

G.T. Dounias
*Edafos SA, Athens, Greece*

S.L. Lazaridou
*Hydroexigiantiki Consulting Engineers, Athens, Greece*

Z.R. Papachatzaki
*Public Power Corporation SA, Athens, Greece*

ABSTRACT: The paper presents and discusses incidents and failures that have occurred in Greek Dams. The incidents and failures are grouped according to the mechanism that caused the incident. The first and most common mechanism involves internal erosion, mostly in homogeneous embankment dams. There are nine cases reported. In some, the deficiency was treated, and the dams are properly operating. In others, the reservoir is lowered waiting for remedial measures. In a few cases, erosion was not controlled and led to dam breach. The second mechanism involves overtopping of dams. There are seven reported cases. In many cases, the dam was badly damaged and had to be repaired. In some cases, very little or no damage occurred, mainly due to the rockfill of the downstream shoulder. A review of design floods and probable works to enhance spillway capacities are needed. The last group concerns excessive leakage. Four cases are reported, one concerning extreme leakage in the reservoir. The other cases involve excessive leakage through the foundation/abutments which are difficult to treat successfully. Investigations are ongoing to find a solution in each case. Most of the incidents would have been avoided if well-established principles, such as presented by ICOLD Bulletins, had been followed. The need for dam design and construction made by specialists and not by inexperienced personnel is apparent.

*Role of Dams and Reservoirs in a Successful Energy Transition – Boes, Droz & Leroy (Eds)*
*© 2023 The Author(s), ISBN 978-1-032-57668-8*

# A dam-foundation seismic interaction analysis method: Development and first case studies

G. Faggiani & P. Masarati
*Ricerca sul Sistema Energetico – RSE S.p.A., Italy*

ABSTRACT: To accurately and realistically define the response of concrete dams to earthquakes it is often necessary to carry out complex FEM (Finite Element Method) analyses able to consider the three-dimensionality of the problem, the semi-unbounded size of the foundation, the non-linear behaviour of the system and the dynamic interactions of the dam with the foundation and the reservoir. An advanced approach to deal with dam-foundation seismic interaction was recently implemented and tested with the objective of appropriately simulating the propagation of seismic waves in a realistic massed foundation, considering its semi-unbounded extent. The approach, indicated by the acronym SAM-4D (Seismic Advanced Model for Dams), was developed by RSE to overcome the main deficiencies (above all the excessive conservativeness of the results) of traditional and simplified methods. Artificial non-reflecting (or absorbing) boundaries are used to delimit the semi-unbounded foundation and effective earthquake forces, computed with reference to the elastic wave vertical propagation theory, provide the seismic motion: the method can be adopted within the framework of FEM codes able to address all the above-mentioned complexities and allows to reduce the recognized excessive conservativity of the results obtained with simplified traditional approaches (e.g. the massless approach). Pre-processing software tools were developed to automate the calculation and the assignment of the effective earthquake forces to be applied to the artificial boundaries of the foundation. The method, including the pre-processing tools, was validated using theoretical and numerical solutions available in technical literature. Three case studies of dam-reservoir-foundation systems (one 2D case and two 3D cases) were carried out, allowing to verify the reliability of the method for real applications and to highlight the strong conservativity of the traditional analysis approaches with respect to the proposed method.

*Role of Dams and Reservoirs in a Successful Energy Transition – Boes, Droz & Leroy (Eds)*


# Concrete swelling: Studies and pragmatic results. Case study of Cleuson dam

Jonathan Fauriel
*Head Civil Engineering and Environment, Alpiq Ltd, Switzerland*

Andrea Abati
*Projet manager, Gruner Stucky SA, Switzerland*

Milaine Côté
*Supervision of Dams Specialist, Swiss Federal Office of Energy SFOE, Switzerland*

Reynald Berthod
*N3 Expert, Gruner Stucky SA, Switzerland*

ABSTRACT: Concrete swelling affects most of the existing concrete dams in Switzerland. The peculiar case study of the Cleuson dam, a 87 m hollow gravity dam, is proposed.

Irreversible movements of the dam body have been measured almost since the first impoundment and significant cracking in the upper part of the structure has been monitored since 2005. Based on the observation of the cracking and preliminary calculations, an operating restriction of 2 m was decided in 2008. In 2012, further investigations led to characterization of the cracking and determination of the concrete pathology. On this basis, a finite element model was made in 2013, implementing the movements, the cracks and the concrete swelling. The results showed that the structural safety was guaranteed.

In view of the dam body swelling pursuit, new studies were carried out in 2021 and 2022 to estimate the time frame in which the safety of the structure would no longer be guaranteed. Analyses of the probable failure mechanism associated with a damage model showed that a risk of failure should not appear before 30 years.

*Role of Dams and Reservoirs in a Successful Energy Transition – Boes, Droz & Leroy (Eds)*
*© 2023 The Author(s), ISBN 978-1-032-57668-8*

# The safety assessment of buttress, hollow gravity and multiple arch/slab dams. The contribution of numerical modeling

A. Frigerio & M. Colombo
*Ricerca sul Sistema Energetico -RSE S.p.A., Italy*

G. Mazzà
*ITCOLD, Italy*

F. Rogledi & A. Terret
*Consorzio di Bonifica di Piacenza, Italy*

ABSTRACT: An overview of the typical problems concerning buttress, hollow gravity and multiple arch/slab dams is presented, emphasizing the support that advanced numerical modeling can provide for the evaluation of their safety and for the definition of the most effective interventions to guarantee long-term safety conditions.

Even if the construction of this type of dams is nowadays practically abandoned, the problem of assessing their safety remains an important and challenging issue considering the obsolescence of most existing structures of this type and the peculiarities connected with their structural behavior, in particular referring to the effects of aging and seismic loadings.

In the Italian context there are numerous cases of this type of works built between the two World Wars and immediately after the II World War. A total of 37 dams are in operation (40% buttress, 30% hollow gravity, 30% multiple arch/slab dams). The problems connected with these structures are well-known: cracks (e.g., caused by phenomena of thermal origin, differences of the buttresses height, expansive chemical reactions, etc.), degradation associated with environmental conditions (e.g., due to corrosion of reinforcement bars when present), and aging. Other criticalities are related to compliance with current legislation which requires the verification of more stringent conditions not foreseen in the design phase (for instance higher seismic loads).

Some of the above-mentioned problems have been investigated in the frame of Benchmark Workshops proposed by the ICOLD Technical Committee A "*Computational Aspects of Analysis and Design of Dams*", and are analyzed by an Italian Working Group, established by ITCOLD, the Italian National Committee on Large Dams.

Above all, in this paper the application of the eXtended Finite Element Method to evaluate the propagation of cracks in concrete dams and the seismic reassessment of an Italian multiple arch dam are discussed.

*Role of Dams and Reservoirs in a Successful Energy Transition – Boes, Droz & Leroy (Eds)*
*© 2023 The Author(s), ISBN 978-1-032-57668-8*

# Rehabilitation and upgrade of old small dams

Brigitta Gander
*Dam Safety Authority, Canton of Zürich, Switzerland*

ABSTRACT: There are more than 350 small dams in the Canton of Zurich, Switzerland. Some date back to the medieval ages and many were built between 1750 AD and 1900 AD. Today, these small dams are often protected for environmental or historical reasons. As the Federal Act on Water Retaining Facilities came in force in 2010, the supervision of small dams was handed over from the Federal Government to the cantons. The review of the dams by the Dam Safety Authority of the Canton of Zurich showed that hardly any of these small dams meet the requirements of the Federal Act. In this paper the reasons for dam rehabilitation and some examples of cost-effective measures to improve the safety of small dams are discussed. From several upgrading projects, the renewal of the spillway of the Sülibachweiher near Bauma and the retrofitting of the Sternensee above Richterswil are selected as case studies.

*Role of Dams and Reservoirs in a Successful Energy Transition – Boes, Droz & Leroy (Eds)*


# Experimental investigation of the overtopping failure of a zoned embankment dam

M.C. Halso, F.M. Evers, D.F. Vetsch & R.M. Boes
*Laboratory of Hydraulics, Hydrology & Glaciology (VAW), ETH Zurich, Zurich, Switzerland*

ABSTRACT: The failure of dams poses an immense risk to settlements and infrastructure throughout the world. As extreme flood events become more likely in many parts of the world, dams become increasingly vulnerable to overtopping. Earthen embankment dams are made of erodible materials, and are therefore susceptible to erosion by an overtopping flow. Erosion of a dam to the point of uncontrolled outflow constitutes a dam breach. The erosive processes that lead to breaching depend on the type of dam. Zoned earthen dams with a mineral core erode differently than homogeneous earthen dams, due to the core material's resistance to entrainment by the overtopping flow. The process of breaching a zoned dam by overtopping has been scarcely observed at laboratory, field, or prototype scales. In this study, we perform a laboratory experiment to investigate the breaching due to overtopping of a zoned earthen embankment dam. A prototype-scale zoned dam was designed based on multiple large zoned earthen dams in Switzerland, with particular focus on the Jonenbach Dam in Affoltern am Albis. The prototype dam, which includes shell, filter, and core zones, was reduced to laboratory scale for experimentation. The laboratory-scale dam was breached by overtopping, and the failure processes of each zone were observed. The breaching process began with formation of a breach channel on the downstream slope of the dam. The breach channel incised downstream of the core, and expanded laterally due to mass slope failures of shell material. The filter material remained temporarily stable after departure of the supporting shell material, due to effects of apparent cohesion, but gradually failed with mass detachments. Erosion of the shell and filter zones left the core unsupported on the downstream side. The forces of water and soil pressure from upstream gradually became too large for the core to resist, causing the core to bulge, crack, and eventually break. Once the core broke, water and shell material from upstream flowed uncontrolled through the breach. Similar morphodynamic and hydrodynamic processes that occurred in this experiment would also be expected to occur during the overtopping failure of a prototype-scale zoned dam. The resistance of the core to the soil and water pressures can be a valuable output for calibration of statics calculations of core stability. Such calculations could be implemented in parametric or numerical dam breach models, for use by engineers to estimate the timing and resulting discharge of zoned earthen dam breaches.

# Reconstruction of hydrographs of the maximum annual flood event at dam site

Working Group of the Italian National Committee on Large Dams, ITCOLD (*)
*(*) The Working Group was made up of members from the national dam authority, dam owners and universities: F. Santoro & F. Pianigiani (Italian Dam Authority, Rome, Italy), A. Bonafè, L. Ruggeri & D. Feliziani (Enel Green Power Italy, Venice and Terni, Italy), M. Maestri (Alperia, Bolzano, Italy), F. Piras (ENAS, Cagliari, Italy), F. Sainati (Edison, Milan, Italy). P. Claps (Turin Polytechnic, Turin, Italy), A. Brath (Bologna University, Bologna, Italy)*

ABSTRACT: In 2000, the Italian Dam Authority issued a procedure to regulate the measures that should be carried out during each significant flood event, with the aim of obtaining a hydrological database of annual maximum flows.

After 15 years of applying this procedure, the Italian Dam Authority revised it in 2018 and required dam owners to reconstruct the most significant hydrological event of the year and the most significant events of the previous five-year period. They also provided further technical indications to improve and optimize the procedure. However, the experience gained in the last three years has shown that the reconstruction of hydrological events was affected by errors due to tolerance in reservoir level measurements, uncertainties in the measurement of flow rates, and discharge curves that were not always adequately calibrated.

With the aim, shared by the operators and the authority, to increase the database of the maximum annual flows with values that are "reliable hydrological data", ITCOLD constituted the Working Group on the reconstruction of the maximum annual hydrographs. In March 2023 the group published the final report, providing more detailed technical indications for the reconstruction of hydrographs of the maximum annual flood event and defining shared criteria for the identification of cases of exemption from this obligation.

*Role of Dams and Reservoirs in a Successful Energy Transition – Boes, Droz & Leroy (Eds)*


# Experiences from inspections and controls on ageing penstocks of hydropower plants

A. Kager & R. Sadei
*HYDRO SAFETY Engineering d. Kager Armin*

ABSTRACT: In hydropower, the use of steel pipes has proven its worth in the realization of penstocks. Especially at high pressure and large diameters, steel pipes are still unrivalled. However, the longevity of the infrastructure can only be guaranteed if operation is accompanied by periodic checks and professional maintenance. The present technical paper reflects the experience gained during numerous inspections of penstocks over the past 10 years.

*Role of Dams and Reservoirs in a Successful Energy Transition – Boes, Droz & Leroy (Eds)*


# Instable rock cliff at Steinwasser water intake: Immediate and safety measures

A. Koch
*IUB Engineering Ltd., Bern, Switzerland*

D. Bürki
*Kraftwerke Oberhasli Ltd., Innertkirchen, Switzerland*

ABSTRACT: The Steinwasser water intake is a 20 m high dam which was endangered by an instable rock cliff located above its left bank. In 2021 the tachymetric monitoring assessed an acceleration of the deformations. The safety of the Steinwasser dam was assessed using an acting force derived from the crash model provided by geologists. The results showed that the structural safety of the dam in case of an uncontrolled rockslide is not fulfilled, thus immediate measures were planned to excavate the material in stages with controlled blasting. The dam was protected by a 1 m thick Rockfall-X mattress and part of the reservoir near the left dam abutment was filled with gravel material to dampen the impact of blasted rocks. The blastings were implemented in such a way that the detached elements (individual boulders) were no bigger than 1 $m^3$ in volume. Vibrations were measured during the construction works in order to ensure the safety of the dam. Regular measurements and the early implementation of appropriate measures are the key components to ensuring a high level of safety while guaranteeing the operation of the Steinwasser water intake.

*Role of Dams and Reservoirs in a Successful Energy Transition – Boes, Droz & Leroy (Eds)*


# Emergency preparedness planning in Greek dams

S.L. Lazaridou
*Hydroexigiantiki Consulting Engineers, Athens, Greece*

G.T. Dounias & S.C. Sakellariou
*Edafos SA, Athens, Greece*

A.E. Kotsonis, G.P. Kastranta & M.V. Psychogiou
*Hellenic Republic Ministry of Infrastructure and Transport, Athens, Greece*

ABSTRACT: The paper discusses the main procedures for Emergency Preparedness Planning (EPP) in Greece. The Greek Dam Safety Regulation (DSR) came into force in October 2017, and a Dam Administrative Authority (DAA) was formed to manage its application. One of the main obligations of the dam operators is the preparation of an Emergency Preparedness Plan (EPP). The DAA prepared an EPP Standard, to be followed by Dam Owners and Operators. After discussions with the dam community and the Civil Protection Agency, the Standard was finalized in 2021. All dams within the jurisdiction of the DSR must prepare an updated EPP, following the new Standard. The Standard introduced a hazard potential class analysis, altering the categorization of many dams which was initially based on dam visible height and reservoir capacity. Regional Authorities and Private Companies who act as Dam Operators must comply to the standard's procedures and define responsibilities, resources and equipment for preventive and emergency actions. In some cases, Dam Operators adapting to this more demanding Regulatory framework, must proceed to significant organizational changes. The EPP provides extended information and data to the Civil Protection Agencies involved, to assist them identify downstream risks and conduct evacuation plans. The Dam Owner, the dam Operator and the Civil Protection Agencies need to collaborate and establish a solid and effective communication pathway. The initial experience of the Standard's application is briefly discussed, and some challenges faced by Dam Owners and Operators during the EPP implementation are presented.

*Role of Dams and Reservoirs in a Successful Energy Transition – Boes, Droz & Leroy (Eds)*


# Seismic analysis of old embankment dams: Qualification of the Fr-Jp method

N. Lebrun
*EDF-CIH, La Motte-Servolex, France*

M. Jellouli
*ISL, Paris, France*

J.-J. Fry
*J-J Fry Consulting, Bassens, France*

ABSTRACT: Embankment dams built before 1950 can pose safety problems, as they were built with poorly to moderately compacted materials and without respecting the current state of the art. Inexpensive, easy and reliable tools can be useful for the seismic reassessment of these dams, especially when the owners of the dams have limited financial resources, but these tools have to be sophisticated enough to be able to fit the cyclic pore pressure generation and the dynamic performance of the dam. The current simplified analyses unfortunately do not reproduce the pore pressure generation up to liquefaction and its consequence on dynamic behavior, only take into account the 2D characteristics of dams and ignore the loss of resistances with deformation. Thus, the Fr-Jp method, developed thanks to JCOLD data, sought to fill these gaps by aligning itself with the seismic behavior observed on JDEC recordings on Japanese dams. EDF with the participation of ISL is in the process of putting this method under quality assurance. Software manual, scientific notice, user manual and validation notice are being drafted. This paper describes how the method is designed and justified by its calibration fitting the accelerograms measured on the Japanese dams. A couple of applications are described.

RÉSUMÉ: Les barrages en remblai construits avant 1950 peuvent poser des problèmes de sécurité, car ils ont été construits avec des matériaux peu ou moyennement compactés et ils ne respectent pas l'état de l'art actuel. Des outils peu coûteux, faciles d'emploi et fiables peuvent être adaptés pour la réévaluation sismique de ces barrages, en particulier quand leurs propriétaires ont des moyens financiers limités, mais ces outils doivent être suffisamment sophistiqués pour pouvoir reproduire la génération cyclique de pression interstitielle et le comportement dynamique du barrage. Les analyses simplifiées actuelles ne reproduisent malheureusement pas la génération de pression interstitielle jusqu'à la liquéfaction et ses conséquences sur le comportement dynamique, ne prennent en compte que les caractéristiques 2D des barrages et ignorent la perte de résistance avec la déformation. Aussi, la méthode Fr-Jp, développée grâce aux données de JCOLD, a cherché à combler ces lacunes et à reproduire le comportement sismique enregistré sur les barrages japonais par JDEC. EDF, avec la participation d'ISL, est en train de mettre cette méthode sous assurance qualité. Le manuel du logiciel, l'avis scientifique, le manuel de l'utilisateur et l'avis de validation sont en cours de rédaction. Cet article décrit comment la méthode est conçue et justifiée par son étalonnage en fonction des accélérogrammes mesurés sur les barrages japonais. Deux applications sont décrites.

*Role of Dams and Reservoirs in a Successful Energy Transition – Boes, Droz & Leroy (Eds)*
*© 2023 The Author(s), ISBN 978-1-032-57668-8*

# Obturation solutions for dry works on underwater installations

M. Leon
*Hydrokarst-SDEM, Grenoble, France*

ABSTRACT: The obturation of a sluice, bottom outlet or water intake, with subaquatic means, is a way to perform the refurbishment of existing installations and equipment for dams, especially for dry inspections, gate maintenance and reparation/painting of the lining. The obturation avoids the drainage of the full reservoir and all the associated losses: floods management, hydroelectric production, water supply. This can be a solution also if the sluice has no functional maintenance gate or if the refurbishment includes the upstream equipment's (steel lining, trash rack, concreting). The design and construction of these obturators must consider the Civil Works environment (shape and condition), the lack of underwater visibility of the divers, the anchoring system, the sealing and all the logistic for the transportation, handling, erection, drilling and grouting. The efficiency of the watertightness and the transmission of the thrust loads on the existing Civil Works might require the addition of an embedded frame with grouting. Another issue to consider properly is the self-resistance of the sluice against external pressure, directly downstream the obturator. Even if the sluice has a steel liner, its embedding with the concrete structure or its thickness might not be sufficient to avoid a collapse after dewatering the sluice. In this case, stiffener rings or stabilization girders must be added downstream obturation.

RÉSUMÉ: L'obturation des pertuis de barrage par des moyens subaquatiques peut s'avérer nécessaire pour réaliser des travaux de maintenance à sec: inspection, réparation ou remplacement des vannes, travaux de peinture, soudure ou bétonnage. C'est le cas notamment en l'absence de vanne de maintenance fonctionnelle ou si des réparations doivent être menées en amont de celle-ci. Grâce à l'obturation, l'exploitant du barrage peut maintenir la retenue amont à un niveau normal pour éviter les pertes de stockage d'eau, de production hydroélectrique et assurer la continuité de la gestion des crues. La conception et la fabrication de l'obturateur doivent tenir compte d'un ensemble de paramètres dont notamment: la forme et l'état de l'ouvrage béton, le manque de visibilité des scaphandriers pour opérer, les choix techniques pour l'ancrage et l'étanchéité et tous les moyens logistiques pour le transport sur site, la manutention, le montage, le forage des ancrages et si nécessaire, le bétonnage. Pour assurer une bonne étanchéité de l'ensemble d'obturation et transmettre correctement les descentes de charges au génie civil, il peut s'avérer nécessaire d'installer un cadre fixe scellé avec du mortier. Un point à ne pas négliger est la vérification l'auto-résistance du pertuis à la pression extérieure. Même en présence d'un blindage, la stabilité de celui-ci peut être insuffisante car non prévue pour ce cas de charges déséquilibré après la vidange du pertuis. Dans ce cas, des étayages ou des renforts devront être installés en aval de l'obturateur.

*Role of Dams and Reservoirs in a Successful Energy Transition – Boes, Droz & Leroy (Eds)*


# The decommissioning of dams in Italy: The state of the art

P. Manni
*ITCOLD Decommissioning and downgrade of dams' observatory*

G. Mazzà
*Vice Chairman of ITCOLD – Italian National Committee on Large Dams*

ABSTRACT: Large Dams under state jurisdiction in Italy are 530. The mean age of the Italian dams is approximately 65 years. About 60% of the national dams are predominantly for hydroelectric use, 26% for irrigation and 12% for drinking water; the remaining are dedicated to different uses. The main aspects covered in the paper refers to (i) features of the decommissioned dams; (ii) most frequent reasons that determine the decommissioning; (iii) rules and procedures governing the approval of decommissioning projects; (iv) main decommissioning solutions over the past few years.

The publisher will reduce the camera-ready copy to 75%.

RÉSUMÉ: Les grands barrages sous juridiction étatique en Italie sont au nombre de 530. L'âge moyen des barrages italiens est d'environ 65 ans. Environ 60 % des barrages nationaux sont principalement à usage hydroélectrique, 26 % pour l'irrigation et 12 % pour l'eau potable; les autres sont dédiés à différents usages. Les principaux aspects couverts dans le document concernent: (i) les caractéristiques des barrages déclassés; (ii) les raisons les plus fréquentes qui déterminent le déclassement; (iii) les règles et procédures régissant l'approbation des projets de déclassement; (iv) les principales solutions de déclassement au cours des dernières années.

*Role of Dams and Reservoirs in a Successful Energy Transition – Boes, Droz & Leroy (Eds)*
*© 2023 The Author(s), ISBN 978-1-032-57668-8*

# Rehabilitation of the Pàvana Dam in Tuscany (IT) Advantages from the use of building information modelling in the design of a complex hydraulic project

F. Maugliani, A. Piazza, D. Longo, G. Raimondi & R. Sanfilippo
*Lombardi Engineering Ltd., Switzerland*

A. Parisi
*Enel Green Power Italia S.r.l., Italy*

ABSTRACT: Located in Italy, at the border between the two municipalities of Castel di Casio (BO) and Sambuca Pistoiese (PT), the buttress dam of Pàvana is a work of historical and architectural value, serving the Suviana hydroelectric cascade.

The dam, built in the 1920s, is managed by Enel Green Power Italia S.r.l., the major Italian energy production company.

The rehabilitation works of the hundred-year-old dam are required due to an alkali-aggregate reaction (AAR) mainly affecting the central portion of the structure.

The rehabilitation project foresees the upgrade of the central dam portion from a buttress dam to a gravity dam, by means of the demolition of the upper section of the three arches and the two buttresses of the central dam body, and the concreting of the volumes between the existing buttresses (including the demolished buttresses themselves). The project is completed by the construction of a new bottom outlet, in order to overcome the actual limits in operation due to sedimentation.

For the detailed design of this complex structural modification, the Building Information Modelling (BIM) method was used, according to a proposal by the designer, well accepted and confirmed by the owner's technicians: a challenging and ambitious activity that allowed for collaborative and efficient work between the various actors involved thanks to its digital sharing tools and the use of a Common Data Environment for the document management, 3D visualization of the various issues, clash detection, and automatic information management.

The model of the existing dam was reconstructed based on the laser scanning survey of the dam and of the previous definitive design made by 2D drawings developed by the dam owner and from a point cloud survey.

The modelling included the complex demolition and construction phases of the individual dam components and a complete overview of the construction sequences (BIM4D).

The BIM model was used not only for the production of the 2D drawings and of the bill of quantities, but also as a basis for the structural and geotechnical FE analysis as well as the three-dimensional geological model, where the results of the geological investigations were integrated.

RÉSUMÉ: Situé en Italie dans la municipalité de Pàvana (PT), le barrage à contreforts du même nom est un ouvrage de valeur historique et architecturale, desservant la cascade hydroélectrique de Suviana et géré par la société italienne de production d'énergie Enel Green Power Italia S.r.l.. Datant des années 1920, les travaux de réhabilitation sont nécessaires en raison d'une réaction alcali-agrégat (RAG) qui affecte principalement la partie centrale de la structure. Le projet de consolidation se concentre sur la conversion de la section centrale du barrage d'un barrage à contreforts en un barrage-poids et implique la démolition de la partie supérieure des trois arches et des deux contreforts du corps du barrage central, ainsi que le remplissage en béton des espaces entre les contreforts existants (y compris les contreforts démolis eux-mêmes). Le projet comprend également la construction d'une nouvelle décharge de fond, étant donné que la décharge existante est actuellement totalement envasée. Pour la conception exécutive de cette structure plutôt complexe, la méthode Building Information Modeling (BIM) a été choisie: une activité ambitieuse et difficile qui a permis un travail collaboratif et cohérent entre les différents acteurs impliqués grâce à ses outils de partage numérique et à l'utilisation d'un environnement de données commun pour la gestion des documents, la visualisation en 3D des différentes questions et la gestion automatique des informations. Le modèle du barrage existant a été reconstruit sur la base des dessins 2D de la conception finale développés par le propriétaire du barrage et à partir d'un relevé de nuages de points. La modélisation comprenait les phases complexes de démolition et de construction des différents éléments du barrage, ainsi qu'un aperçu complet des séquences de construction (BIM4D). Le modèle BIM a été utilisé non seulement pour la production des dessins 2D et du devis, mais aussi pour l'analyse structurelle et géotechnique par éléments finis ainsi que pour le modèle géologique tridimensionnel, dans lequel les résultats des études géologiques ont été intégrés.

*Role of Dams and Reservoirs in a Successful Energy Transition – Boes, Droz & Leroy (Eds)*


# Potential Failure Mode Analysis (PFMA) to deal with ageing and climate change affecting dams

Philippe Méan
*Ing.Civil EPFL, Ph.D UCD, former Director Energie Ouest Suisse, Switzerland*

Thomas Bryant
*M.Engineering CEngineering MICE, Senior Dam Engineer, Gruner Stucky SA, Switzerland*

ABSTRACT: this article highlights the value of performing a qualitative risk assessment in the form of a *Potential Failure Mode Analysis* (PFMA) to address uncertainties related to Climate Change and Ageing. As they constitute new and overlapping threats with numerous potential dam safety issues, engineers should consider the joint impact of Ageing and Climate Change on dam safety with a specific risk analysis. The dam engineers must here content themselves with a qualitative assessment since the non-stationary processes of Ageing and Climate Change are not easily reducible to statistical analysis. The identification of the potential situations that place the dam at highest risk must be carried out simultaneously and systematically to all dam components. Partial investigations can actually be misleading and result in underestimated risks and missing the potential interdependencies. An explicit method is required to identify globally new dam vulnerabilities potentially induced by Ageing and Climate Change and to verify their influence on the failure modes previously recognized at construction time. An expert's implicit choice of dangerous situations should be avoided as arbitrary. The regular completion of a PFMA on the occasion of a periodic "Comprehensive Safety Inspection" (every 5-year in Switzerland) and the inclusion of its results in the corresponding "Dam Safety Assessment Report" would address uncertainties related to Climate Change and Ageing on dam safety until the next periodic formal Safety Inspection. This would dynamically increase stakeholder's risk awareness and address the best strategy to enhance the dam preparedness and resilience in a holistic framework and potentially contribute to help focus research efforts. It is the authors' belief that *Potential failure mode analysis* (PFMA) could be in practice implemented more widely in Switzerland and in several other EU countries, as a complement to their Standard-based dam regulation, in a period of rapid change and increasing uncertainty calling for regular reviews based on the latest observations.

RESUME: cet article souligne l'importance d'effectuer une évaluation qualitative des risques sous la forme d'une *Analyse des Modes Potentiels de Défaillance* (AMPD) afin de tenir compte des incertitudes liées aux Changement Climatique et au Vieillissement sur la sécurité des barrages. Une méthode explicite est nécessaire pour identifier les nouveaux risques affectant potentiellement les barrages et pour vérifier leur influence conjointe sur les modes de défaillance précédemment reconnus au moment de la construction. On doit ici se contenter d'une évaluation qualitative car les processus non stationnaires du Vieillissement et du Changement Climatique ne se réduisent pas facilement à l'analyse statistique. L'exécution périodique d'une AMPD à l'occasion d'un «Examen périodique approfondi de la sécurité » (tous les 5 ans en Suisse) et l'inclusion des résultats dans le «Rapport d'expertise sur la sécurité du barrage » correspondant permettrait d'adresser les incertitudes liées au Changement Climatique et au Vieillissement jusqu'à la prochaine inspection périodique. Cela permettrait d'accroître de façon dynamique la sensibilisation des intervenants au risque, de mettre en œuvre la meilleure stratégie pour améliorer la résilience des barrages et d'orienter les éventuel efforts de recherche. Dans une période de changements rapides et d'incertitude croissante qui exigent des analyses régulières fondées sur les dernières observations, les auteurs sont d'avis que *l'Analyse des Modes Potentiels de Défaillance* (AMPD) pourrait être appliquée plus largement en Suisse comme dans plusieurs autres pays de l'UE, en complément de leur réglementation fondée sur les Normes.

# GRIMONIT (GroundRiskMonitor) – an early warning system for difficult measurement conditions

E. Meier & I. Gutiérrez
*Edi Meier + Partner AG, Winterthur, Switzerland*

M. Büchel
*Marc Büchel Media – ocaholic AG, Freienstein, Switzerland*

ABSTRACT: In the context of hazard prevention, it is of central importance to continuously monitor deformations on bridges, dams, buildings or in underground mining. Equally challenging is the permanent monitoring of forested landslide areas that pose a risk to railroad tracks or roads. "GRIMONIT" (Ground Risk Monitor) is a fully automated hydrostatic measurement system, which is a further development of the LAS-Meter, developed in cooperation with the ETH. In this paper a measurement with the LAS-Meter is compared with GRIMONIT-measurements. The time span between the two measurements is 11 years. The task was to document the subsidence at the landfill site "Zingel".

RESUMEN: En el contexto de la prevencion de riesgos es de suma importancia la monitorización continuada de deformaciones en puentes, presas, edificios o mineria subterranea. Igualmente desafiante es el monitoreo permanente de zonas de deslizamientos boscosas que suponen un riesgo para lineas de ferrocaril o carreteras. GRIMONIT es un sistema de medición hidrostático completamente automatizado, resultante del perfeccionamiento del LAS-Meter y desarrollado en cooperación con la ETH. En este artículo se compara una medición llevada a cabo con el LAS-Meter con otra usando el GRIMONIT, realizada en el mismo lugar 11 años depués. El objetivo era documentar una posible subsidencia en la base del vertedero "Zingel".

*Role of Dams and Reservoirs in a Successful Energy Transition – Boes, Droz & Leroy (Eds)*
*© 2023 The Author(s), ISBN 978-1-032-57668-8*

# Arch dams: A new methodology to analyse the sliding stability between the dam and the foundation

X. Molin, C. Jouy, S. Delmas, P. Anthiniac, G. Milesi & C. Noret
*Tractebel Engineering, France*

ABSTRACT: Arch dams transfer the hydrostatic forces to the foundation, by arch effect. Due to their hyperstaticity, they usually benefit from a significant safety factor. The potential sliding at the dam/foundation contact is a failure mode highlighted by most of the international recommendations. Recent FEM analyses carried out on operational dams confirm that thermal cases can be sensitive conditions, especially in summer. Back analyses require generally to introduce high shear strength on the dam/foundation interface, considered unrealistic, to demonstrate the past good dam behaviour. Tractebel has developed a methodology which takes into account the irreversible displacements which can occur in this area, and the redistribution of internal forces. It has been inspired by the latest CFBR recommendations on arch dams. This methodology was applied to three existing thin arch dams, located in France, in wide valleys. Back-analyses calculations were performed on historical loads. This method provides more realistic shear strength values (friction angle of 55 degrees or less) to justify the observed good behaviour of the dams, while larger values, greater than 70 degrees, were required if no displacement was accepted. Finally, the possibility of accepting irreversible displacements at the dam/foundation interface opens the discussion of the peak and residual shear strengths. Definition of acceptable irreversible displacements is discussed.

RÉSUMÉ: Les barrages voûtes transfèrent les forces hydrostatiques à la fondation par effet voûte. Ils bénéficient généralement d'un facteur de sécurité important. La stabilité au glissement du contact barrage fondation du barrage est un mode de rupture retenu par la plupart des recommandations internationales. Les modèles EF récents réalisés sur d'anciens barrages confirment que les cas thermiques peuvent être des conditions sensibles, surtout en été. Les rétroanalyses nécessitent d'introduire une résistance élevée au cisaillement, considérée comme irréaliste, pour démontrer le bon comportement du barrage observé jusqu'à présent. Tractebel a développé une méthodologie qui prend en compte les déplacements irréversibles qui peuvent se produire dans cette zone, et la redistribution des efforts internes. Il s'inspire des dernières recommandations du CFBR sur les barrages voûtes. Cette méthodologie a été appliquée à trois barrages voûtes minces existants, situés en France, en vallée large. Des calculs de rétro-analyse ont été effectués sur les charges historiques. Cette méthode fournit des valeurs de résistance au cisaillement plus réalistes (angle de frottement de 55 degrés ou moins) pour justifier le bon comportement observé des barrages, tandis que des valeurs plus importantes, supérieures à 70 degrés, sont requises si aucun déplacement n'est accepté. Enfin le principe des déplacements irréversibles ouvre la discussion sur les résistances au cisaillement de pic et résiduelle. La définition des déplacements irréversibles acceptables est discutée.

*Role of Dams and Reservoirs in a Successful Energy Transition – Boes, Droz & Leroy (Eds)*


# Spitallamm Arch Dam – Challenges faced for replacing the existing Old Dam

E.A. Odermatt & A. Wohnlich
*Gruner Stucky Ltd, Renens, Switzerland*

ABSTRACT: The 113 m high Spitallamm arch-gravity dam, together with the Seeuferegg gravity dam, form the Grimsel reservoir with a live storage of 94 Million m$^3$ at an altitude between 1'800 and 1'900 masl in the Swiss Alps (Canton of Bern). The dam was erected from 1926 to 1932. It was at that time one of the highest concrete dams in Europe. The dam is set in a narrow gorge cut mainly in Aare granite and granodiorite, both rocks of high strength and low deformability.

After almost one century of operation, the dam has to be replaced, following damages at the upstream dam face. The 113 m high new Spitallamm arch dam is currently being constructed downstream of the old arch-gravity dam. The dam completion and first impounding are scheduled in 2025. The old arch-gravity dam will remain in place and be flooded in the new reservoir.

The paper explains the context for such replacement, addressing first the reasons why the old dam has to be decommissioned, then explaining how it was decided to go for a completely new dam located right downstream of the existing dam, and finally presenting selected interesting technical issues.

*Role of Dams and Reservoirs in a Successful Energy Transition – Boes, Droz & Leroy (Eds)*


# Concrete dams upgrading using IV-RCC

F. Ortega
*RCC dams expert, FOSCE Consulting Engineers, Germany/Spain*

ABSTRACT: Immersion Vibrated Roller Compacted Concrete (IV-RCC) is a mass concrete for dam construction. A special mix design has been recently developed for the construction of new concrete dams with the aim of providing best quality to RCC dams, similar or better than the traditional conventional vibrated concrete (CVC). The key advantage of this new type of concrete is that it can be placed and consolidated either like RCC or like traditional CVC. The decision on which methodology should be used for its application can be decided on site, at the point of placement. In this context, we see an opportunity for making good use of this material in the raising, upgrading and renewal works of existing concrete dams in Europe and elsewhere. The paper describes specific characteristics of the materials and mixes required for IV-RCC, as well as some of the specific construction aspects and advantages. The in-situ performance based on existing completed projects is also presented in the paper. Particular attention is paid to specific high-quality mixes as those required, for example, for concrete arch dams.

# The driving force of AAR – An in-situ proof

B. Otto & R. Senti
*Axpo Power AG, Baden, Switzerland*

ABSTRACT: The paper is focused on the AAR behavior of two dams in the Swiss Alps. Instead of measuring and analyzing the crest deformation, numerous strain distributions across the entire dam body were measured by means of extensometers (rocmeters) and sliding micrometers. The measurements were taken periodically during a year cycle and are running since 4 to 5 years. The analysis of the data showed a strong variation of the yearly AAR strain growth over the dam cross section. While the upper crest area swells 200 – 250 µm/m/year, the dam concrete in 20 m depth from the top of the dam swells only by 50 µm/m/year. On the upstream side the effect of the cold water is very clearly seen, the swelling rate drops within 5 meters of height by a factor of 5. These strong strain gradients can only be explained by the dominant influence of temperature on the swelling process. This finding is strongly confirmed by the AAR during a year cycle. Swelling occurs only in the warm period, in winter the expansion is ceased completely. The swelling rate is an exponential function of the concrete temperature and unique for each dam depending on the type and size of aggregate.

RÉSUMÉ: L'article se concentre sur le comportement RAG de deux barrages dans les Alpes suisses. Au lieu de mesurer et d'analyser les déformations du couronnement, de nombreuses distributions de tension sur l'ensemble du corps du barrage ont été mesurées à l'aide d'extensomètres (rocmètres) et de micromètres coulissants. Les mesures ont été prises périodiquement au cours d'un cycle annuel depuis 4 à 5 ans. L'analyse des données a montré une forte variation de la croissance annuelle de la déformation RAG sur la section transversale du barrage. Alors que la zone de la crête supérieure se gonfle de 200 à 250 µm /m/an, le béton du barrage à une profondeur de 20 m à partir du couronnement du barrage ne se gonfle que de 50 µm /m/an. Du côté amont, l'effet de l'eau froide est très clairement visible, le taux de gonflement diminue d'un facteur 5 à 5 mètres. Ces gradients importants de déformation ne peuvent s'expliquer que par l'influence dominante de la température sur le processus de gonflement. Cette constatation est fortement confirmée par la RAG au cours d'un cycle annuel. Le gonflement ne se produit que pendant la période chaude, en hiver l'expansion est complètement arrêtée. Le taux de gonflement est une fonction exponentielle de la température du béton et est unique pour chaque barrage en fonction du type et de la taille des agrégats.

# Effects of wall roughness on low-level outlet performance

S. Pagliara & B. Hohermuth
*Laboratory of Hydraulics, Hydrology and Glaciology (VAW), ETH Zurich, Zurich, Switzerland*

S. Felder
*Water Research Laboratory, School of Civil and Environmental Engineering, UNSW Sydney, Australia*

R.M. Boes
*Laboratory of Hydraulics, Hydrology and Glaciology (VAW), ETH Zurich, Zurich, Switzerland*

ABSTRACT: Reservoir dams play a key role in modern society, water resources management and economy. Low-level outlets (LLOs) represent important safety structures for regulating the water level in the reservoir, and for its fast drawdown in case of scheduled maintenance or emergency situations. The flow in LLO tunnels is characterized by high velocities and turbulence levels, leading to air entrainment and transport. This results in sub-atmospheric air pressures, which may induce and aggravate serious issues such as gate vibration and cavitation. An adequate flow aeration via an air vent can mitigate these problems and is key to good performance. While many studies focused on the effects of hydraulic parameters, tunnel geometry and air vent design on the air demand of LLOs, the influence of the tunnel wall roughness is still unclear. To this end, physical model tests were carried out to investigate the effects of invert, soffit, and sidewall roughness on the LLO performance, for various combinations of gate opening, energy head at the gate and air vent properties. The roughness modelled in this study represents unlined rock, and it was implemented by attaching expanded aluminum plates to the inner sides of the outlet tunnel. Air velocities in the air vent were measured to estimate the air demand, and pressures along the tunnel were recorded to assess cavitation potential. For rough wall conditions, both the air demand and the cavitation risk were found to increase compared to the smooth tunnel conditions (i.e., acrylic-made invert, walls, and soffit in the model). In conclusion, the study represents a preliminary analysis of the effects of LLO tunnel roughness on air demand and cavitation occurrence, and future research is needed to enable a more quantitative assessment of the differences in air demand between model and real-world prototypes.

*Role of Dams and Reservoirs in a Successful Energy Transition – Boes, Droz & Leroy (Eds)*
*© 2023 The Author(s), ISBN 978-1-032-57668-8*

# Safety of embankment dams in the case of upgrading the existing tailings storage facilities

L. Petkovski, F. Panovska & S. Mitovski
*Faculty of Civil Engineering, Ss. Cyril and Methodius University in Skopje, Skopje, North Macedonia*

ABSTRACT: The embankments over tailings dams and waste lagoons or the upgrade at the existing tailings storage facilities, from stability aspects of a heterogenic geo environment, has many similarities with the tailings dams with upstream construction method. These earth-fill structures are susceptible to liquefaction during static and dynamic (cyclic) loading and therefore they are civil engineering structures with the highest stability risk. The need to provide an additional volume for depositing tailings material, necessary for the regular operation of mines in conditions of spatial limitation, actualizes the upgrade of the tailings storage facilities. This upgrade is characterized by detailed geotechnical in-situ investigations and sophisticated structural analyses, which are illustrated by the results of the stability analyses (in static and dynamic conditions) of a dry stacking embankment above the tailings storage facility Sasa no. 2, Makedonska Kamenica, Republic of North Macedonia.

*Role of Dams and Reservoirs in a Successful Energy Transition – Boes, Droz & Leroy (Eds)*


# First rehabilitation measures of the Biópio dam, Angola

C.J.C. Pontes
*CAB, Luanda, Angola*

P. Afonso
*PRODEL, EP, Luanda, Angola*

ABSTRACT: The Biópio dam is a 19 m high structure, concluded in 1956 on the Catumbela river, about 50 km from the town of Lobito (Angola), being its main purpose the production of energy. The dam is essentially composed by the bottom discharge, on the left bank, and seven spill weir blocks with thin piers, supporting 15 m high automatic tilt gates and a roadway deck, with a total length of 156 m. Inspections carried out in 2007 and 2009 revealed, besides a large deterioration of the dam, appurtenant works and respective equipment, the development of large cavities in some piers, near the support of the gates, which allow the reservoir water to enter into and put under pressure the drainage gallery. As a first step for the full rehabilitation and upgrading of the dam, the owner (the National Energy Enterprise of Angola- ENE EP), with the support of the Energies of Portugal – EDP International, considered of interest the first repair of the above referred to cavities. This paper reports on the major aspects of these works.

*Role of Dams and Reservoirs in a Successful Energy Transition – Boes, Droz & Leroy (Eds)*


# Dams in Angola, reconstruction of the Matala dam

C.J.C. Pontes & P. Portugal
*CAB-Angolan Committee of Dams, Luanda, Angola*

ABSTRACT: Since 2008, after signing the peace agreements in 2002, Angola has been carrying out several dam constructions and rehabilitation projects, with the aim of developing the country's economy and serving its population better, in agricultural irrigation, potable water supply and power generation. Most Angolan dams were affected, directly or indirectly, during the armed conflict that took place for about 30 years. Some suffered from sabotage actions, or from the absence of adequate maintenance and/or rehabilitation actions for long periods. Further, in the case of the Matala dam, the concrete has developed alkalis reactions that boosted concrete degradation.

RÉSUMÉ: Depuis 2008, après la signature des accords de paix en 2002, l'Angola a mené à bien plusieurs projets de construction et de réhabilitation de barrages dans le but de développer l'économie du pays et de mieux servir sa population, dans les domaines de l'irrigation agricole, de l'approvisionnement en eau potable et de la sécurité alimentaire, ainsi que la production d'électricité. La plupart des barrages angolais ont été touchés, directement ou indirectement, par le conflit armé qui a duré environ 30 ans. Certains ont souffert d'actions de sabotage ou du manque d'entretien et/ou de réhabilitation adéquates pendant longues périodes. Pour ce qui concerne le cas particulier du barrage de Matala, le béton a développé des réactions alcali-agrégats qui ont accéléré la dégradation du béton.

*Role of Dams and Reservoirs in a Successful Energy Transition – Boes, Droz & Leroy (Eds)*
*© 2023 The Author(s), ISBN 978-1-032-57668-8*

# Estimation of settlement in earth and rockfill dams using artificial intelligence technique

Mohammad Rashidi
*Stantec, Civil Engineer, Denver, USA*

Kwestan Salimi
*Stantec, Geotechnical Engineer, Denver, USA*

Sam Abbaszadeh
*Stantec, Senior Civil Engineer, Denver, USA*

ABSTRACT: The impermissible settlement is one of the most important reasons for the failures of earth and rockfill dams. An appropriate estimation of settlement after construction is required to evaluate the performance of the dam and to inform dam design engineers of any possible problem. This study was designed to apply artificial intelligent methods to predict settlement after constructing central core rockfill dams. Attempts were made in this research to prepare models for predicting settlement of these dams using the information of 50 different central core earth and rockfill dams all over the world using Genetic Programming (GP) methods. The results indicated that prediction of settlement based on the single parameter of dam height cannot be accurate and other parameters such as stiffness and shear strength properties of materials are effective in dam settlement. Based on the sensitivity analysis, parameters such as height of dam, modulus of elasticity and unit weight of core, and modulus of elasticity and internal friction angle of rockfill materials have the highest influence on the settlement of the dams and were considered as the input parameters. The results obtained from comparing the artificial intelligent method and empirical relationships showed that GP results were more appropriate tools to solve the problems with complex mechanisms and several effective factors, such as prediction of settlement of dams. The new developed models in this study are ready to be applied as a robust predictor tool for monitoring and safety evaluation of earth and rockfill dams in Europe.

*Role of Dams and Reservoirs in a Successful Energy Transition – Boes, Droz & Leroy (Eds)*


# Structural health monitoring of large dams using GNSS and HSCT-FE models. Swelling effect detection

M. Rodrigues, J.N. Lima & S. Oliveira
*Laboratório Nacional de Engenharia Civil, Lisbon, Portugal*

J. Proença
*CERIS - Instituto Superior Técnico IST-ID, Lisbon, Portugal*

ABSTRACT: The use of Global Navigation Satellite System (GNSS) displacement monitoring technology enables permanent and remote monitoring of large dams. The validation of GNSS displacements monitoring system installed in Cabril dam since 2016 has been accomplished by comparing the GNSS measurements not only with triangulation and plumb line methods, but also with Finite Element Models (FEM) and Hydrostatic-Seasonal-Creep-other Time effects (HSCT) models for effects separation. Analysis of a 7-year monitoring period has demonstrated the precise and reliable nature of this method, as it successfully detected the development of pathological behaviour trends associated with concrete swelling reactions.

*Role of Dams and Reservoirs in a Successful Energy Transition – Boes, Droz & Leroy (Eds)*
*© 2023 The Author(s), ISBN 978-1-032-57668-8*

# Dynamic behavior of exposed geomembrane systems in pressure waterways

S. Vorlet, R.P. Seixas & G. De Cesare
*Platform of Hydraulic Constructions (PL-LCH), Ecole Polytechnique Fédérale de Lausanne (EPFL), Lausanne, Switzerland*

ABSTRACT: Geomembrane systems have been used in dams and reservoirs as rehabilitation technology since several decades and are now used worldwide. They act as impervious layer to prevent and mitigate water leakage and damage to structures. They meet the needs of many challenges faced by aging dams by improving their performance and lifespan, enhancing their resilience and sustainability. More recently, their application was extended to pressure waterways and surge shafts. A Finite Element model is developed to investigate the dynamic behavior of a framed hyperelastic geomembrane specimen for enhanced application in pressure waterways accounting for the dynamic behavior of the geomembrane system and fluid-structure interactions in the frequency domain. The nonlinear constitutive behavior of the geomembrane is modeled by the Mooney-Rivlin equation. The effect of water is considered by the added mass approach for the modal characteristics of the geomembrane. The damping is included as Rayleigh damping. Results show that the modal characteristics of the geomembrane are strongly influenced by the material nonlinear constitutive behavior. The first natural frequencies of the hyperelastic geomembrane specimen are found at low frequencies in vacuum. The natural frequencies also strongly increase with the increase of pre-tension in vacuum. In the presence of water, the variation of the natural frequency with pre-tension is highly reduced. The increase in hydrostatic pressure tends to moderately increase the natural frequencies of the specimen. Finally, the damping ratio has almost no influence on the natural frequencies.

*Role of Dams and Reservoirs in a Successful Energy Transition – Boes, Droz & Leroy (Eds)*
*© 2023 The Author(s), ISBN 978-1-032-57668-8*

# The Rigoso project. Two old masonry dams to be recovered

L. Serra
*Waterways, Rome, Italy*

G. Gatto & F. Bisci
*Studio Speri, Rome, Italy*

M. Rebuschi
*Frosio Next, Brescia, Italy*

F. Fornari & L. Dal Canto
*Enel Green Power, Rome, Italy*

ABSTRACT: The Rigoso project includes two dams (Lake Verde and Lake Ballano), built in 1907-8 for hydroelectric purposes, in stone masonry and based on morainic debris. Over the years, interventions have been made to strengthen the dam sections, necessary for the water losses of the foundation. Currently the two works are in disuse due to drastic reservoir limitations imposed by the Italian agency for large dams. The need to fix them definitively and to partially recover their hydroelectric use prompted the Operator, Enel Green Power, to plan their substantial restructuring, which involves reducing the height, making the structures safe with interventions to consolidate facing and foundations, the creation of an adequate drainage system. As a result of the reduction in height, the project is configured as a partial decommissioning. The design had to face not insignificant challenges, given that the constraint of implanting a new project over a century old works, based on permeable loose soils, requires greater accuracy than those required in the design of new works. A further challenge is added to the technical requirements, in fact the two works are located in a Regional Park, and therefore any demolished material must be classified and recovered in the works.

*Role of Dams and Reservoirs in a Successful Energy Transition – Boes, Droz & Leroy (Eds)*


# Numerical modelling of the Pian Telessio dam affected by AAR

R. Stucchi & E. Catalano
*Lombardi Engineering Ltd., Giubiasco, Ticino, Switzerland*

ABSTRACT: The Pian Telessio dam, an 80 m high arch-gravity dam located in Northern Italy, is suffering from the effects of an alkali-aggregate reaction (AAR) since the second half of the 70s. The concrete expansion due to the AAR causes an upstream drift that reached approximately 60 mm in 2008. In 2008 rehabilitation works by means of vertical slot cuttings in the upper half of the dam were performed to reduce the effects of the concrete expansion. The rehabilitation works allowed a minor recovery of the upstream drift in the order of 5-10 mm, lower than expected.

The structural safety of the dam has been re-evaluated by means of a 3D numerical model, including the effect of the AAR expansion. The implemented AAR model, specifically developed for this analysis, accounts for several aspects affecting the swelling reaction: reaction kinetics, state of stress and temperature. The results of the calibration were remarkably satisfactory showing an agreement with several field measurements: displacements, joints opening, crack pattern and state of stress.

*Role of Dams and Reservoirs in a Successful Energy Transition – Boes, Droz & Leroy (Eds)*
*© 2023 The Author(s), ISBN 978-1-032-57668-8*

# Flood protection levees – from an existing portfolio of old structures to safe and reliable protection systems

Rémy Tourment
*INRAE, France*

Adrian Rushworth
*Environment Agency, UK*

Jonathan Simm
*HR Wallingford, UK*

Robert Slomp
*Rijkswaterstaat, The Netherlands*

Malcolm Barker
*GHD, Australia*

David Bouma
*Tonkin + Taylor Ltd., New Zealand*

Noah Vroman
*US Army Corps of Engineers Levee Safety Center – USA*

ABSTRACT: In many countries, levees and flood defences raise (or have raised until recently) different issues associated to their ageing and often too to a lack of proper management during long periods of time. This leads to uncertain safety and performance, and sometimes catastrophic consequences. In this paper, based on an international confrontation of experience and lessons learned, we present the most important of these issues and a framework to better organize the management of these structures: identification of existing portfolios of structures, organizing them into consistent systems, the need for a high-level policy, for a proper local management and for technical guidance.

RÉSUMÉ: Dans de nombreux pays, les digues et autres ouvrages de protection contre les inondations soulèvent (ou ont soulevé jusqu'à récemment) différents problèmes liés à leur vieillissement et souvent aussi à un manque de gestion appropriée pendant de longues périodes. Cela conduit à une sécurité et des performances incertaines, et parfois des conséquences catastrophiques. Dans cet article, basé sur une confrontation internationale de retours d'expérience, nous présentons les plus importants de ces problèmes et un cadre pour mieux organiser la gestion de ces ouvrages: le recensement des structures existantes, les organiser en systèmes cohérents, la nécessité d'une politique de haut niveau, d'une bonne gestion locale et d'un corpus technique de référence.

*Role of Dams and Reservoirs in a Successful Energy Transition – Boes, Droz & Leroy (Eds)*


# Software tool for progressive dam breach outflow estimation

D.F. Vetsch, M.C. Halso, L. Seidelmann & R.M. Boes
*Laboratory of Hydraulics, Hydrology and Glaciology, ETH Zurich, Zurich, Switzerland*

ABSTRACT: Hazard assessment due to a dam failure is an important task in risk management. Assessment of the related hazard potential is based on a dam break analysis, a multi-step workflow in which the first step includes the analysis of possible dam failure and estimation of the outflow hydrograph. For that purpose, various approaches with different level of details with regard to modelled physical processes are available. Among the simplest approaches are parameter models, which are often used due to their straightforward and efficient application. The open-source software BASEbreach provides a suite of parameter models, developed particularly for the estimation of the outflow discharge resulting from progressive dam failure. We illustrate the software capabilities by comparing different approaches for instantaneous and progressive dam failure. Further, we show that the maximum breach discharge may be overestimated or underestimated depending on the approach and the situation. Estimation of the maximum breach discharge is always associated with great uncertainties. Hence, the BASEbreach software provides access to relevant parameters for local sensitivity analysis. Also, Monte-Carlo simulations in combination with the Peter parameter models are available for uncertainty quantification. The software is a valuable tool for engineers and practitioners to estimate the potential breach outflow from progressive embankment dam failure.

# Ageing and life-span of dams

M. Wieland
*ICOLD Committee on Seismic Aspects of Dam Design, Honorary Member of ICOLD, Dam Consultant, Dietikon and AFRY Switzerland, Zurich, Switzerland*

ABSTRACT: Ageing is the main factor governing the life-span of any engineering structure. In the context of this paper the term ageing is used to describe the decrease in safety of the dam body with time. The factors to be considered include all time-dependent physical processes and changes in safety criteria that govern the safety assessment of dams. Long life-spans can only be reached, if the dam satisfies all safety criteria during its long life, which implies that the owner or operator provides take care of a dam and provide adequate maintenance.

*Role of Dams and Reservoirs in a Successful Energy Transition – Boes, Droz & Leroy (Eds)*
 *ISBN 978-1-032-57668-8*

# Hongrin arch dams – Rehabilitation works of the central artificial gravity abutment

A. Wohnlich
*Gruner Stucky Ltd, Renens, Switzerland*

R. Leroy
*Alpiq Ltd, Lausanne, Switzerland*

ABSTRACT: The Hongrin hydroelectric pumped storage scheme is located in Switzerland (Canton of Vaud); it was constructed at the end of the 1960s and presents two concrete arch dams 123 m high (northern dam) and 90 m high (southern dam). Both arch dams are connected by means of an artificial gravity abutment located on the central hill dividing both valleys.

The scheme is now more than 50 years old and over the decades, damages have materialized on the central artificial gravity abutment; namely, marked concrete cracking appeared along some horizontal construction joints, probably caused by insufficient concrete curing at the time of the construction (thermal cracking due to heat hydration). Over the years, such cracking pattern caused slow irreversible downstream displacements of the artificial gravity abutment in the range of 8 to 10 mm, which tend to increase over time.

The paper presents and discusses how the case was handled, starting for the monitored behaviour of both dams, the observation of damages in the artificial gravity abutment especially after the 2015 heatwave, the engineering studies considering a 3D model of both dams and the simulation of 15 years of operation, leading to the strengthening project consisting in the implementation of a total of approx. 50'000 kN vertical prestressed anchors in the artificial gravity abutment. The rehabilitation works took place in 2018. As a result, favorable results in terms of structural behavior are being observed.

*Role of Dams and Reservoirs in a Successful Energy Transition – Boes, Droz & Leroy (Eds)*
*© 2023 The Author(s), ISBN 978-1-032-57668-8*

# Dealing with aging dams on the Drava River in Slovenia

P. Žvanut
*Slovenian National Building and Civil Engineering Institute, Ljubljana, Slovenia*

ABSTRACT: There are eight hydropower plants (HPPs) on the Slovenian part of the Drava River. Six HPPs (Dravograd, Vuzenica, Vuhred, Ožbalt, Fala and Mariborski otok) are situated directly in the river course, while the other two HPPs (Zlatoličje and Formin) are located in derivation channels of the river. For this purpose, ten concrete gravity dams were built between 1918 and 1978 with a structural height between 17 and 54 m. Due to the great age of these dams, it is necessary to monitor the condition of the dam structures and their surroundings even more carefully, which enables appropriate action in the case of identified deficiencies. These activities are carried out through regular annual technical observations, which include the monitoring of the deformations of the dams and the filtration of groundwater in the wider area of these structures, as well as regular accurate visual inspections. In order to gain a better insight into the condition of dams, some measurements (mainly hydrostatic and partly hydrodynamic) have already been automated. Recently, drones have also been included in the monitoring of dams, which allow insight into the condition of dam structures and their areas of influence, even in hard-to-reach or inaccessible places. Data obtained through technical observation of dams are also used for numerical analyses of dams, namely for the calibration of numerical models for calculating the static and the dynamic safety of dams. The results of the surveillance of these dams showed that, mainly due to the aging of the dam structures and also due to extraordinary events (i.e. flood and equipment failure), the renovations of the dams were necessary. Due to the aging of the dams, the renovation of the oldest dam on the Drava River in Slovenia (Fala Dam) began as early as 1987, which was followed by the renovation of seven more dams (Dravograd, Vuzenica, Vuhred, Ožbalt, Mariborski otok, Melje and Zlatoličje) in the following years. The remaining two dams (Markovci and Formin) still need to be renovated; the renovation of the Markovci Dam is currently underway (it should be completed in 2026), while the start of renovation of the youngest dam on the Drava River in Slovenia (Formin Dam) is scheduled for 2024. The renovation included the revitalization of the mechanical, electrical and structural parts of the dams. After the complete renovations of the dams, the surveillance systems of the dams were updated. Two dams were also renovated due to extraordinary events. The rehabilitation of the Formin Dam - due to damage after floods - was performed between 2013 and 2014, while the rehabilitation of the Mariborski otok Dam - due to damage following equipment failure - was carried out between 2018 and 2020. These dams needed to be properly renovated so that they continue to serve their purpose well in the coming decades.

# Author index